RAND NATIONAL DEFENSE RESEARCH INSTITUTE

T0146369

SEXUAL ASSAULT AND SEXUAL HARASSMENT IN THE U.S. MILITARY

Annex to Volume 3. Tabular Results from the
2014 RAND Military Workplace Study for
Coast Guard Service Members

Andrew R. Morral, Kristie L. Gore, Terry L. Schell, editors

Prepared for the DoD Sexual Assault Prevention and Response Office

For more information on this publication, visit www.rand.org/t/RR870z5

Library of Congress Cataloging-in-Publication Data is available for this publication.
ISBN: 978-0-8330-9076-8

Support RAND
Make a tax-deductible charitable contribution at
www.rand.org/giving/contribute

www.rand.org

The 2014 RAND Military Workplace Study Team

Principal Investigators
Andrew R. Morral, Ph.D.
Kristie L. Gore, Ph.D.

Instrument Design
Lisa Jaycox, Ph.D., team lead
Terry Schell, Ph.D.
Coreen Farris, Ph.D.
Dean Kilpatrick, Ph.D.*
Amy Street, Ph.D.*
Terri Tanielian, M.A.*

Study Design and Analysis
Terry Schell, Ph.D., team lead
Bonnie Ghosh-Dastidar, Ph.D.
Craig Martin, M.A.
Q Burkhart, M.S.
Robin Beckman, M.P.H.
Megan Mathews, M.A.
Marc Elliott, Ph.D.

Project Management
Kayla M. Williams, M.A.
Caroline Epley, M.P.A.
Amy Grace Donohue, M.P.P.

Survey Coordination
Jennifer Hawes-Dawson

Westat Survey Group
Shelley Perry, Ph.D., team lead
Wayne Hintze, M.S.
John Rauch
Bryan Davis
Lena Watkins
Richard Sigman, M.S.
Michael Hornbostel, M.S.

Project Communications
Steve Kistler
Jeffrey Hiday
Barbara Bicksler, M.P.P.

Scientific Advisory Board

Major General John Altenburg, Esq. (USA, ret.)
Captain Thomas A. Grieger, M.D. (USN, ret.)
Dean Kilpatrick, Ph.D.
Laura Miller, Ph.D.
Amy Street, Ph.D.
Roger Tourangeau, Ph.D.

David Cantor, Ph.D.
Colonel Dawn Hankins, USAF
Roderick Little, Ph.D.
Sharon Smith, Ph.D.
Terri Tanielian, M.A.
Veronica Venture, J.D.

* Three members of the Scientific Advisory Board were so extensively involved in the development of the survey instrument that we list them here as full Instrument Design team members.

Preface

The Department of Defense (DoD) has assessed service member experiences with sexual assault and harassment since at least 1996, when Public Law 104-201 first required a survey of the "gender relations climate" experienced by active-component forces. Since 2002, four "Workplace and Gender Relations Surveys," as they are known in 10 U.S.C. 481, have been conducted with active-component forces (in 2002, 2006, 2010, and 2012). DoD conducted reserve-component versions of this survey in 2004, 2008, and 2012.

The results of the 2012 survey suggested that more than 26,000 service members in the active component had experienced *unwanted sexual contacts* in the prior year, an estimate that received widespread public attention and concern. In press reports and congressional inquiries, questions were raised about the validity of the estimate, about what "unwanted sexual contact" included, and about whether the survey had been conducted properly. Because of these questions, some members of Congress urged DoD to seek an independent assessment of the number of service members who experienced sexual assault or sexual harassment.

The Sexual Assault Prevention and Response Office within the Office of the Secretary of Defense selected the RAND Corporation to provide a new and independent evaluation of sexual assault, sexual harassment, and gender discrimination across the military. As such, DoD asked the RAND research team to redesign the approach used in previous DoD surveys, if changes would improve the accuracy and validity of the survey results for estimating the prevalence of sexual crimes and violations. In the summer of 2014, RAND fielded a new survey, called the RAND Military Workplace Study.

This Annex to Volume 3 contains detailed tabular results for the Coast Guard active component. The complete series that collectively describes the study methodology and its findings includes the following reports:

- *Sexual Assault and Sexual Harassment in the U.S. Military: Top-Line Estimates for Active-Duty Service Members from the 2014 RAND Military Workplace Study*
- *Sexual Assault and Sexual Harassment in the U.S. Military: Top-Line Estimates for Active-Duty Coast Guard Members from the 2014 RAND Military Workplace Study*

- *Sexual Assault and Sexual Harassment in the U.S. Military: Volume 1. Design of the 2014 RAND Military Workplace Study*
- *Sexual Assault and Sexual Harassment in the U.S. Military: Volume 2. Estimates for Department of Defense Service Members from the 2014 RAND Military Workplace Study*
- *Sexual Assault and Sexual Harassment in the U.S. Military: Annex to Volume 2. Tabular Results from the 2014 RAND Military Workplace Study for Department of Defense Service Members*
- *Sexual Assault and Sexual Harassment in the U.S. Military: Volume 3. Estimates for Coast Guard Service Members from the 2014 RAND Military Workplace Study*
- *Sexual Assault and Sexual Harassment in the U.S. Military: Annex to Volume 3. Tabular Results from the 2014 RAND Military Workplace Study for Coast Guard Service Members*
- *Sexual Assault and Sexual Harassment in the U.S. Military: Volume 4. Investigations of Potential Bias in Estimates from 2014 RAND Military Workplace Study.*

These reports are available online at www.rand.org/surveys/rmws.html.

This research was conducted within the Forces and Resources Policy Center of the RAND National Defense Research Institute, a federally funded research and development center sponsored by the Office of the Secretary of Defense, the Joint Staff, the Unified Combatant Commands, the Navy, the Marine Corps, the defense agencies, and the defense Intelligence Community.

For more information on the Forces and Resources Policy Center, see http://www.rand.org/nsrd/ndri/centers/frp.html or contact the director (contact information is provided on the web page).

Contents

Abbreviations

DoD	Department of Defense
JAG	Judge Advocate General
NCO	noncommissioned officer
NR	not reportable
SAPR VA	Sexual Assault Prevention and Response Victim Advocate
SARC	sexual assault response coordinator
TA	temporary additional duty
TDY	temporary duty
VA	victim advocate

Statistical Analysis and Reporting Conventions Used in This Report

The statistical analyses presented in Volume 3 and this annex employ statistical procedures designed to reduce the likelihood of drawing inappropriate conclusions or compromising the privacy of respondents.

All survey respondents were assured in the survey Privacy Statement that our reports would not include analyses conducted on subsets smaller than 15 respondents. For that reason, the annexes do not include sample statistics (including confidence intervals) computed within groups smaller than 15 unweighted respondents. If such a cell appears in a table, the point estimates and its confidence intervals are replaced with "NR" (not reportable).

These annex tables contain estimates that vary dramatically in their statistical precision. This occurs because some percentages are estimated on more than 100,000 respondents, while others are estimated on small subsamples. To reduce the likelihood of misinterpretations, highly imprecise percentages are not reported. Specifically, percentages estimated with a margin of error greater than 15 percentage points are replaced with "NR," where the margin of error is defined as the larger half-width of the confidence interval. In such cases, the confidence intervals, which can be measured precisely, are still presented to communicate the range of percentages that are consistent with the data. That is, a confidence interval of 45 percent to 97 percent, for instance, indicates that we can have 95-percent confidence that the correct value for the population lies within that range. Conversely, we have 95-percent confidence that the true value for the population is not lower than 45 percent or higher than 97 percent.

Editor's note: Below its title, each table contains the survey question(s) on which the results are based; the entire survey instrument is contained in Volume 1 of this series.

Sexual Assault in the U.S. Coast Guard: Detailed Results

A.1. Percentage of members who experienced a sexual assault in the past year

Table A.1
Percentage of members who experienced a sexual assault in the past year, by gender and pay grade

4 level pay grade grouping	Total	Men	Women
Total	0.69% (0.46–1.00)	0.29% (0.09–0.71)	2.97% (2.25–3.83)
E1–E4	0.83% (0.49–1.31)	0.00% (0.00-0.45)	4.01% (2.81–5.53)
E5–E9	0.63% (0.27–1.25)	0.46% (0.12–1.22)	2.18% (1.10–3.84)
O1–O3	0.63% (0.24–1.35)	0.00% (0.00–0.96)	2.70% (1.33–4.83)
O4–O6	0.61% (0.05–2.48)	0.55% (0.01–3.01)	0.96% (0.12–3.42)

NOTE: Includes estimates for active-component Coast Guard members. 95-percent confidence intervals for each estimate are included in parentheses.

Too few warrant officers were included in the sample to break them out as a separate pay grade. For the purposes of this table, warrant officers have been included with the E5–E9 category.

A.2. Percentage of members who experienced any type of unwanted event of an abusive, humiliating, or sexual nature

Table A.2
Percentage of members who experienced any type of unwanted event of an abusive, humiliating, or sexual nature, by gender and pay grade

4 level pay grade grouping	Total	Men	Women
Total	1.25% (0.91–1.67)	0.85% (0.49–1.36)	3.53% (2.76–4.44)
E1–E4	1.60% (1.05–2.32)	0.79% (0.27–1.77)	4.70% (3.41–6.30)
E5–E9	1.20% (0.69–1.95)	1.05% (0.51–1.91)	2.52% (1.38–4.18)
O1–O3	0.81% (0.35–1.6)	0.00% (0.00–0.96)	3.49% (1.84–5.9)
O4–O6	0.68% (0.08–2.47)	0.55% (0.01–3.01)	1.44% (0.30–4.13)

NOTE: Includes estimates for active-component Coast Guard members. 95-percent confidence intervals for each estimate are included in parentheses.

Too few warrant officers were included in the sample to break them out as a separate pay grade. For the purposes of this table, warrant officers have been included with the E5–E9 category.

A.3. Percentage of members who experienced a sexual assault in their lifetime

Table A.3
Percentage of members who experienced a sexual assault in their lifetime, by gender and pay grade

4 level pay grade grouping	Total	Men	Women
Total	4.50% (4.01–5.03)	1.85% (1.39–2.41)	19.57% (17.94–21.28)
E1–E4	4.13% (3.33–5.06)	1.19% (1.02–2.42)	15.40% (13.02–18.03)
E5–E9	3.49% (2.85–4.24)	1.61% (1.02–2.42)	20.31% (17.53–23.32)
O1–O3	8.45% (6.62–10.58)	3.70% (1.96–6.30)	24.05% (19.90–28.59)
O4–O6	8.01% (5.96–10.50)	4.01% (2.09–6.91)	32.22% (26.13–38.79)

NOTE: Includes estimates for active-component Coast Guard members. 95-percent confidence intervals for each estimate are included in parentheses.

Too few warrant officers were included in the sample to break them out as a separate pay grade. For the purposes of this table, warrant officers have been included with the E5–E9 category.

A.4. Percentage of members who experienced a sexual assault prior to joining the military

Table A.4
Percentage of members who experienced a sexual assault prior to joining the military, by gender and pay grade

4 level pay grade grouping	Total	Men	Women
Total	1.66% (1.36–2.01)	0.58% (0.31–0.98)	7.83% (6.74–9.02)
E1–E4	1.81% (1.28–2.47)	0.36% (0.08–1.05)	7.36% (5.64–9.40)
E5–E9	1.26% (0.83–1.81)	0.54% (0.18–1.26)	7.65% (5.94–9.67)
O1–O3	2.42% (1.55–3.59)	1.04% (0.31–2.54)	6.94% (4.72–9.79)
O4–O6	2.92% (1.78–4.47)	1.12% (0.30–2.85)	13.77% (9.39–19.23)

NOTE: Includes estimates for active-component Coast Guard members. 95-percent confidence intervals for each estimate are included in parentheses.

Too few warrant officers were included in the sample to break them out as a separate pay grade. For the purposes of this table, warrant officers have been included with the E5–E9 category.

A.5. Percentage of members who experienced a sexual assault since joining the military

Table A.5

Percentage of members who experienced a sexual assault since joining the military, by gender and pay grade

4 level pay grade grouping	Total	Men	Women
Total	3.83 (3.37–4.34)	1.61% (1.18–2.16)	16.45% (14.94–18.05)
E1–E4	3.29% (2.57–4.14)	1.02% (0.43–2.05)	11.98% (9.87–14.34)
E5–E9	3.21% (2.58–3.94)	1.53% (0.95–2.32)	18.22% (15.53–21.15)
O1–O3	7.20% (5.51–9.21)	3.06% (1.47–5.57)	20.83% (16.89–25.21)
O4–O6	6.16% (4.38–8.40)	2.90% (1.27–5.61)	25.90% (20.38–32.04)

NOTE: Includes estimates for active-component Coast Guard members. 95-percent confidence intervals for each estimate are included in parentheses.

Too few warrant officers were included in the sample to break them out as a separate pay grade. For the purposes of this table, warrant officers have been included with the E5–E9 category.

A.6. Number of different sexual assaults experienced in the past year among members who experienced a sexual assault

Table A.6.a
Number of different sexual assaults experienced in the past year among members who experienced a sexual assault, by gender

SAFU1: Please give your best estimate of how many different times (on how many separate occasions) during the past 12 months, you had these unwanted experiences?

	Total	Men	Women
1 time since [X Date]	NR (19.74–49.79)	NR (NR)	50.94% (36.98–64.78)
2 times since [X Date]	17.41% (7.40–32.40)	NR (NR)	17.68% (8.94–29.86)
3 times since [X Date]	6.39% (2.03–14.56)	NR (NR)	8.47% (2.60–19.37)
4 times since [X Date]	3.48% (0.65–10.25)	NR (NR)	3.76% (0.45–12.99)
5 or more times since [X Date]	NR (3.66–54.19)	NR (NR)	3.74% (0.64–11.38)
More than once, but not sure the number of times it happened since [X Date]	NR (5.70–37.75)	NR (NR)	NR (5.54–31.51)

NOTE: Includes estimates for active-component Coast Guard members. 95-percent confidence intervals for each estimate are included in parentheses.

NR = Not reportable.

Table A.6.b
Number of different sexual assaults experienced in the past year among members who experienced a sexual assault, by service

SAFU1: Please give your best estimate of how many different times (on how many separate occasions) during the past 12 months, you had these unwanted experiences?

	Total DoD	Army	Navy	Air Force	Marine Corps	Coast Guard
1 time since [X Date]	33.87% (29.38–38.59)	37.18% (30.53–44.20)	29.09% (20.49–38.95)	45.25% (39.75–50.84)	28.28% (17.63–41.05)	NR (19.74–49.79)
2 times since [X Date]	16.51% (13.54–19.83)	17.12% (12.85–22.12)	15.79% (10.05–23.12)	21.70% (16.90–27.14)	12.34% (6.94–19.78)	17.41% (7.40–32.40)
3 times since [X Date]	10.73% (8.21–13.69)	10.16% (5.85–16.11)	11.34% (7.44–16.33)	9.95% (6.54–14.34)	11.19% (4.38–22.35)	6.39% (2.03–14.56)
4 times since [X Date]	4.37% (2.11–7.89)	2.53% (1.09–4.95)	3.47% (1.71–6.21)	3.71% (1.94–6.37)	NR (1.34–37.48)	3.48% (0.65–10.25)
5 or more times since [X Date]	15.58% (10.62–21.71)	13.14% (7.37–21.07)	20.89% (10.47–35.13)	10.66% (6.62–16.01)	12.03% (4.69–23.96)	NR (3.66–54.19)
More than once, but not sure the number of times it happened since [X Date]	18.95% (14.84–23.64)	19.87% (13.17–28.10)	19.43% (13.75–26.20)	8.73% (6.35–11.64)	NR (9.45–45.77)	NR (5.70–37.75)

NOTE: Includes estimates for active-component DoD and Coast Guard members. 95-percent confidence intervals for each estimate are included in parentheses.

NR = Not reportable.

Table A.6.c
Number of different sexual assaults experienced in the past year among members who experienced a sexual assault, by pay grade

SAFU1: Please give your best estimate of how many different times (on how many separate occasions) during the past 12 months, you had these unwanted experiences?

	Total	E1–E4	E5–E9	O1–O3	O4–O6
1 time since [X Date]	NR (19.74–49.79)	NR (29.83–64.75)	NR (6.79–44.31)	NR (NR)	NR (NR)
2 times since [X Date]	17.41% (7.40–32.40)	NR (9.30–38.44)	NR (1.09–42.95)	NR (NR)	NR (NR)
3 times since [X Date]	6.39% (2.03–14.56)	NR (2.08–27.26)	NR (0.42–21.84)	NR (NR)	NR (NR)
4 times since [X Date]	3.48% (0.65–10.25)	NR (0.86–21.68)	2.26% (0.01–16.71)	NR (NR)	NR (NR)
5 or more times since [X Date]	NR (3.66–54.19)	3.41% (0.10–17.13)	NR (2.29–78.53)	NR (NR)	NR (NR)
More than once, but not sure the number of times it happened since [X Date]	NR (5.70–37.75)	NR (2.78–28.42)	NR (5.89–63.34)	NR (NR)	NR (NR)

NOTE: Includes estimates for active-component Coast Guard members. 95-percent confidence intervals for each estimate are included in parentheses.

Too few warrant officers were included in the sample to break them out as a separate pay grade. For the purposes of this table, warrant officers have been included with the E5–E9 category.

NR = Not reportable.

A.7. Number of offenders in the single or most serious assault among members who experienced a sexual assault in the past year

Table A.7.a
Number of offenders in the single or most serious assault among members who experienced a sexual assault in the past year, by gender

Derived variable combining:

for multiple events: SAFU2: Were all of these events done by the same person? for single events: SAFU5: How many people did this to you?

	Total	Men	Women
Single offender	NR (32.38–73.32)	NR (NR)	NR (52.26–80.60)
More than one offender	NR (23.43–65.60)	NR (NR)	NR (15.31–42.86)
Not sure	3.07% (0.33–11.09)	NR (NR)	4.86% (0.58–16.56)

NOTE: Includes estimates for active-component Coast Guard members. 95-percent confidence intervals for each estimate are included in parentheses.

NR = Not reportable.

Table A.7.b
Number of offenders in the single or most serious assault among members who experienced a sexual assault in the past year, by service

Derived variable combining:

for multiple events: SAFU2: Were all of these events done by the same person? for single events: SAFU5: How many people did this to you?

	Total DoD	Army	Navy	Air Force	Marine Corps	Coast Guard
Single offender	55.16% (49.62–60.60)	59.13% (50.85–67.04)	51.49% (41.00–61.89)	65.58% (59.62–71.21)	NR (29.90–62.89)	NR (32.38–73.32)
More than one offender	41.62% (36.15–47.24)	35.27% (27.69–43.43)	46.41% (35.97–57.09)	32.92% (27.30–38.93)	NR (35.35–68.57)	NR (23.43–65.60)
Not sure	3.22% (1.51–5.95)	5.60% (1.59–13.55)	2.09% (0.84–4.29)	1.49% (0.64–2.92)	1.78% (0.34–5.28)	3.07% (0.33–11.09)

NOTE: Includes estimates for active-component DoD and Coast Guard members. 95-percent confidence intervals for each estimate are included in parentheses.

NR = Not reportable.

Table A.7.c

Number of offenders in the single or most serious assault among members who experienced a sexual assault in the past year, by pay grade

Derived variable combining:

for multiple events: SAFU2: Were all of these events done by the same person? for single events: SAFU5: How many people did this to you?

	Total	E1–E4	E5–E9	O1–O3	O4–O6
Single offender	NR (32.38–73.32)	NR (46.76–81.47)	NR (14.48–77.87)	NR (NR)	NR (NR)
More than one offender	NR (23.43–65.60)	NR (12.22–43.89)	NR (22.13–85.52)	NR (NR)	NR (NR)
Not sure	3.07% (0.33–11.09)	NR (1.13–27.21)	NR (0.00–38.25)	NR (NR)	NR (NR)

NOTE: Includes estimates for active-component Coast Guard members. 95-percent confidence intervals for each estimate are included in parentheses.

Too few warrant officers were included in the sample to break them out as a separate pay grade. For the purposes of this table, warrant officers have been included with the E5–E9 category.

NR = Not reportable.

A.8. Type of single or most serious assault among members who experienced a sexual assault in the past year

Table A.8.a
Type of single or most serious assault among members who experienced a sexual assault in the past year, by gender

Derived variable:

For respondents with a single assault, classification is based on answers to SA1–SA6, PF items, and OB items as described in the report. For respondents with multiple assaults, classification is based on what happened in the most serious assault (responses to SAFU3 items: Which of the following experiences happened during the event you chose as the worst or most serious?), scored hierarchically for mutually exclusive categories.

	Total	Men	Women
Penetrative	NR (28.11–67.59)	NR (NR)	41.28% (28.65–54.81)
Contact	NR (29.19–67.46)	NR (NR)	52.01% (38.30–65.50)
Attempted penetrative	4.31% (0.31–17.19)	NR (NR)	NR (0.55–24.96)

NOTE: Includes estimates for active-component Coast Guard members. 95-percent confidence intervals for each estimate are included in parentheses.

NR = Not reportable.

Table A.8.b
Type of single or most serious assault among members who experienced a sexual assault in the past year, by service

Derived variable:

For respondents with a single assault, classification is based on answers to SA1–SA6, PF items, and OB items as described in the report. For respondents with multiple assaults, classification is based on what happened in the most serious assault (responses to SAFU3 items: Which of the following experiences happened during the event you chose as the worst or most serious?), scored hierarchically for mutually exclusive categories.

	Total DoD	Army	Navy	Air Force	Marine Corps	Coast Guard
Penetrative	36.63% (31.54–41.95)	36.33% (29.12–44.02)	33.34% (23.88–43.89)	35.53% (30.75–40.54)	NR (29.67–63.45)	NR (28.11–67.59)
Contact	60.90% (55.62–66.01)	61.64% (53.97–68.90)	63.45% (53.07–72.99)	61.65% (56.53–66.59)	NR (35.29–68.99)	NR (29.19–67.46)
Attempted penetrative	2.47% (1.83–3.25)	2.03% (1.17–3.27)	3.21% (1.86–5.13)	2.82% (1.68–4.41)	1.42% (0.32–3.97)	4.31% (0.31–17.19)

NOTE: Includes estimates for active-component DoD and Coast Guard members. 95-percent confidence intervals for each estimate are included in parentheses.

NR = Not reportable.

Table A.8.c
Type of single or most serious assault among members who experienced a sexual assault in the past year, by pay grade

Derived variable:

For respondents with a single assault, classification is based on answers to SA1–SA6, PF items, and OB items as described in the report. For respondents with multiple assaults, classification is based on what happened in the most serious assault (responses to SAFU3 items: Which of the following experiences happened during the event you chose as the worst or most serious?), scored hierarchically for mutually exclusive categories.

	Total	E1–E4	E5–E9	O1–O3	O4–O6
Penetrative	NR (28.11–67.59)	NR (31.57–65.49)	NR (18.38–84.00)	NR (NR)	NR (NR)
Contact	NR (29.19–67.46)	NR (33.89–67.82)	NR (12.28–73.02)	NR (NR)	NR (NR)
Attempted penetrative	4.31% (0.31–17.19)	0.67% (0.00–8.82)	NR (0.50–33.46)	NR (NR)	NR (NR)

NOTE: Includes estimates for active-component Coast Guard members. 95-percent confidence intervals for each estimate are included in parentheses.

Too few warrant officers were included in the sample to break them out as a separate pay grade. For the purposes of this table, warrant officers have been included with the E5–E9 category.

NR = Not reportable.

A.9. Gender of offender(s) in the single or most serious assault among members who experienced a sexual assault in the past year

Table A.9.a
Gender of offender(s) in the single or most serious assault among members who experienced a sexual assault in the past year, by gender

Derived variable combining across two items:

For single offender: SAFU6a: Was this person... For more than one offender: SAFU6b: Were these people...

	Total	Men	Women
Man or men only	NR (41.83–91.33)	NR (NR)	92.81% (82.52–98.03)
Woman or women only	NR (1.93–25.88)	NR (NR)	7.19% (1.97–17.48)
A mix of men and women	NR (2.25–55.19)	NR (NR)	0.00% (0.00–5.71)

NOTE: Includes estimates for active-component Coast Guard members. 95-percent confidence intervals for each estimate are included in parentheses.

Some response options that appear in Table A.9.b do not appear here because they were not selected by any Coast Guard respondent.

NR = Not reportable.

Table A.9.b
Gender of offender(s) in the single or most serious assault among members who experienced a sexual assault in the past year, by service

Derived variable combining across two items:

For single offender: SAFU6a: Was this person... For more than one offender: SAFU6b: Were these people...

	Total DoD	Army	Navy	Air Force	Marine Corps	Coast Guard
Man or men only	79.17% (74.21–83.56)	78.54% (70.90–84.97)	79.93% (71.33–86.89)	81.43% (74.69–87.02)	NR (55.24–91.33)	NR (41.83–91.33)
Woman or women only	14.89% (10.86–19.72)	17.90% (11.61–25.77)	12.19% (6.61–20.03)	12.18% (7.30–18.69)	NR (4.14–40.42)	NR (1.93–25.88)
A mix of men and women	5.72% (3.81–8.19)	3.45% (2.11–5.31)	7.63% (3.86–13.27)	6.20% (3.22–10.62)	5.71% (0.86–17.86)	NR (2.25–55.19)
Not sure	0.23% (0.07–0.54)	0.10% (0.00–0.69)	0.25% (0.01–1.14)	0.19% (0.00–1.08)	0.50% (0.01–2.71)	NR (NR)

NOTE: Includes estimates for active-component DoD and Coast Guard members. 95-percent confidence intervals for each estimate are included in parentheses.

NR = Not reportable.

Table A.9.c
Gender of offender(s) in the single or most serious assault among members who experienced a sexual assault in the past year, by pay grade

Derived variable combining across two items:

For single offender: SAFU6a: Was this person... For more than one offender: SAFU6b: Were these people...

	Total	E1–E4	E5–E9	O1–O3	O4–O6
Man or men only	NR (41.83–91.33)	NR (73.01–97.69)	NR (20.01–93.03)	NR (NR)	NR (NR)
Woman or women only	NR (1.93–25.88)	NR (2.31–26.99)	NR (0.00–38.25)	NR (NR)	NR (NR)
A mix of men and women	NR (2.25–55.19)	0.00% (0.00–9.64)	NR (6.97–79.99)	NR (NR)	NR (NR)

NOTE: Includes estimates for active-component Coast Guard members. 95-percent confidence intervals for each estimate are included in parentheses.

Too few warrant officers were included in the sample to break them out as a separate pay grade. For the purposes of this table, warrant officers have been included with the E5–E9 category.

Some response options that appear in Table A.9.b do not appear here because they were not selected by any Coast Guard respondent.

NR = Not reportable.

A.10. Percentage of members who knew the offender(s) in the single or most serious assault among members who experienced a sexual assault in the past year

Table A.10.a
Percentage of members who knew the offender(s) in the single or most serious assault among members who experienced a sexual assault in the past year, by gender

SAFU7: At the time of the event, did you know or had you previously met the person/any of the people who did this to you?

Total	Men	Women
NR	NR	91.76%
(47.47–97.26)	(NR)	(82.12–97.17)

NOTE: Includes estimates for active-component Coast Guard members. 95-percent confidence intervals for each estimate are included in parentheses.

NR = Not reportable.

Table A.10.b
Percentage of members who knew the offender(s) in the single or most serious assault among members who experienced a sexual assault in the past year, by service

SAFU7: At the time of the event, did you know or had you previously met the person/any of the people who did this to you?

Total DoD	Army	Navy	Air Force	Marine Corps	Coast Guard
88.71%	85.05%	92.99%	86.36%	88.68%	NR
(85.22–91.61)	(77.35–90.93)	(88.12–96.30)	(80.80–90.80)	(77.88–95.39)	(47.47–97.26)

NOTE: Includes estimates for active-component DoD and Coast Guard members. 95-percent confidence intervals for each estimate are included in parentheses.

NR = Not reportable.

Table A.10.c
Percentage of members who knew the offender(s) in the single or most serious assault among members who experienced a sexual assault in the past year, by pay grade

SAFU7: At the time of the event, did you know or had you previously met the person/any of the people who did this to you?

Total	E1–E4	E5–E9	O1–O3	O4–O6
NR	NR	NR	NR	NR
(47.47–97.26)	(76.02–98.44)	(21.50–97.61)	(NR)	(NR)

NOTE: Includes estimates for active-component Coast Guard members. 95-percent confidence intervals for each estimate are included in parentheses.

Too few warrant officers were included in the sample to break them out as a separate pay grade. For the purposes of this table, warrant officers have been included with the E5–E9 category.

NR = Not reportable.

A.11. Relationship of offender(s) to victim in the single or most serious assault among members who experienced a sexual assault in the past year and knew the offender(s)

Table A.11.a

Relationship of offender(s) to victim in the single or most serious assault among members who experienced a sexual assault in the past year and knew the offender(s), by gender

SAFU8a–f: At the time of the event, was the person/were the people who did this to you...

	Total	Men	Women
Your spouse?	1.67% (0.09–7.57)	NR (NR)	2.32% (0.20–9.13)
Your boyfriend or girlfriend?	3.58% (0.69–10.45)	NR (NR)	4.98% (1.14–13.31)
Someone you had divorced or broken up with?	2.66% (0.31–9.43)	NR (NR)	3.70% (0.45–12.73)
A friend or acquaintance?	NR (30.05–62.52)	NR (NR)	51.82% (37.09–66.33)
None of the above.	NR (30.21–64.11)	NR (NR)	NR (24.35–53.77)

NOTE: Includes estimates for active-component Coast Guard members. 95-percent confidence intervals for each estimate are included in parentheses.

Some response options that appear in Table A.11.b do not appear here because they were not selected by any Coast Guard respondent.

NR = Not reportable.

Table A.11.b
Relationship of offender(s) to victim in the single or most serious assault among members who experienced a sexual assault in the past year and knew the offender(s), by service

SAFU8a–f: At the time of the event, was the person/were the people who did this to you...

	Total DoD	Army	Navy	Air Force	Marine Corps	Coast Guard
Your spouse?	1.74% (1.12–2.57)	2.10% (0.84–4.32)	1.22% (0.45–2.63)	2.99% (1.52–5.26)	1.26% (0.19–4.15)	1.67% (0.09–7.57)
Your boyfriend or girlfriend?	2.90% (2.07–3.95)	2.77% (1.45–4.75)	2.84% (1.47–4.94)	3.24% (1.92–5.11)	3.08% (1.09–6.73)	3.58% (0.69–10.45)
Someone you had divorced or broken up with?	3.04% (0.89–7.38)	1.46% (0.70–2.68)	1.00% (0.32–2.31)	3.17% (1.69–5.38)	NR (1.06–41.34)	2.66% (0.31–9.43)
A friend or acquaintance?	56.91% (51.06–62.63)	53.02% (45.06–60.87)	59.89% (48.47–70.57)	62.06% (56.16–67.72)	NR (34.52–72.18)	NR (30.05–62.52)
Someone who you have a child with (your child's mother or father)?	0.48% (0.19–1.00)	0.08% (0.00–0.70)	0.58% (0.10–1.80)	0.46% (0.08–1.47)	1.17% (0.16–4.01)	0.00% (0.00–7.90)
A family member or relative?	0.20% (0.05–0.54)	0.06% (0.00–0.66)	0.07% (0.00–0.88)	0.42% (0.06–1.41)	0.66% (0.02–3.67)	0.00% (0.00–7.90)
None of the above.	37.65% (32.12–43.44)	42.10% (34.45–50.05)	37.06% (26.44–48.67)	30.68% (25.05–36.77)	NR (18.96–53.02)	NR (30.21–64.11)

NOTE: Includes estimates for active-component DoD and Coast Guard members. 95-percent confidence intervals for each estimate are included in parentheses.

NR = Not reportable.

Table A.11.c
Relationship of offender(s) to victim in the single or most serious assault among members who experienced a sexual assault in the past year and knew the offender(s), by pay grade

SAFU8a–f: At the time of the event, was the person/were the people who did this to you...

	Total	E1–E4	E5–E9	O1–O3	O4–O6
Your spouse?	1.67% (0.09–7.57)	0.00% (0.00–9.97)	NR (0.00–17.53)	NR (NR)	NR (NR)
Your boyfriend or girlfriend?	3.58% (0.69–10.45)	3.23% (0.09–16.27)	NR (0.00–17.53)	NR (NR)	NR (NR)
Someone you had divorced or broken up with?	2.66% (0.31–9.43)	3.34% (0.10–16.74)	NR (0.00–26.92)	NR (NR)	NR (NR)
A friend or acquaintance?	NR (30.05–62.52)	NR (40.24–75.97)	NR (17.42–72.40)	NR (NR)	NR (NR)
None of the above.	NR (30.21–64.11)	NR (18.71–53.19)	NR (25.89–81.37)	NR (NR)	NR (NR)

NOTE: Includes estimates for active-component Coast Guard members. 95-percent confidence intervals for each estimate are included in parentheses.

Too few warrant officers were included in the sample to break them out as a separate pay grade. For the purposes of this table, warrant officers have been included with the E5–E9 category.

Some response options that appear in Table A.11.b do not appear here because they were not selected by any Coast Guard respondent.

NR = Not reportable.

A.12. Identity of offender(s) in the single or most serious assault among members who experienced a sexual assault in the past year

Table A.12.a
Identity of offender(s) in the single or most serious assault among members who experienced a sexual assault in the past year, by gender

SAFU8g–k: At the time of the event, was the person/were the people who did this to you...

	Total	Men	Women
Someone in the military?			
Yes	NR (65.34–91.23)	NR (NR)	77.15% (64.56–86.96)
No	16.40% (6.71–31.24)	NR (NR)	18.11% (9.72–29.50)
Do not know	3.00% (0.38–10.26)	NR (NR)	4.74% (0.67–15.35)
Civilian employee(s) or contractor(s) working for the military?			
Yes	NR (1.05–55.32)	NR (NR)	2.54% (0.27–9.32)
No	NR (45.23–96.56)	NR (NR)	89.63% (78.38–96.24)
Do not know	4.78% (1.16–12.50)	NR (NR)	7.83% (2.13–19.00)
Person(s) in the local community?			
Yes	NR (12.29–59.31)	NR (NR)	21.15% (11.69–33.59)
No	NR (38.77–84.52)	NR (NR)	73.31% (59.79–84.25)
Do not know	3.41% (0.53–10.84)	NR (NR)	5.54% (0.95–16.55)
A foreign national?			
Yes	NR (2.44–56.36)	NR (NR)	0.00% (0.00–6.13)
No	NR (42.68–93.25)	NR (NR)	89.52% (77.14–96.53)
Do not know	6.41% (1.87–15.20)	NR (NR)	10.48% (3.47–22.86)

NOTE: Includes estimates for active-component Coast Guard members. 95-percent confidence intervals for each estimate are included in parentheses.

NR = Not reportable.

Table A.12.b
Identity of offender(s) in the single or most serious assault among members who experienced a sexual assault in the past year, by service

SAFU8g–k: At the time of the event, was the person/were the people who did this to you...

	Total DoD	Army	Navy	Air Force	Marine Corps	Coast Guard
Someone in the military?						
Yes	85.00%	83.99%	89.59%	77.99%	NR	NR
	(80.90–88.52)	(76.80–89.66)	(84.79–93.28)	(72.16–83.09)	(60.08–94.55)	(65.34–91.23)
No	10.30%	9.21%	6.72%	18.13%	NR	16.40%
	(7.41–13.84)	(6.03–13.32)	(3.86–10.73)	(13.17–24.01)	(3.39–38.75)	(6.71–31.24)
Do not know	4.70%	6.80%	3.69%	3.89%	2.80%	3.00%
	(2.72–7.50)	(2.38–14.70)	(1.76–6.75)	(2.40–5.92)	(0.90–6.50)	(0.38–10.26)
Civilian employee(s) or contractor(s) working for the military?						
Yes	8.90%	9.94%	6.37%	10.83%	11.21%	NR
	(6.41–11.95)	(5.39–16.41)	(3.71–10.08)	(6.83–16.09)	(3.53–24.93)	(1.05–55.32)
No	84.23%	82.26%	90.37%	83.63%	NR	NR
	(79.47–88.27)	(73.65–88.99)	(85.79–93.86)	(78.23–88.14)	(50.81–89.60)	(45.23–96.56)
Do not know	6.87%	7.80%	3.26%	5.54%	NR	4.78%
	(3.73–11.42)	(2.97–16.04)	(1.45–6.23)	(3.45–8.37)	(2.49–42.24)	(1.16–12.50)
Person(s) in the local community?						
Yes	19.23%	21.03%	13.02%	29.80%	21.57%	NR
	(15.94–22.87)	(15.58–27.38)	(8.68–18.52)	(24.15–35.95)	(11.05–35.74)	(12.29–59.31)
No	75.93%	72.37%	84.31%	66.02%	71.83%	NR
	(71.65–79.87)	(64.43–79.40)	(78.26–89.23)	(59.89–71.78)	(56.87–83.93)	(38.77–84.52)
Do not know	4.83%	6.59%	2.66%	4.17%	6.61%	3.41%
	(2.77–7.77)	(2.11–14.95)	(1.03–5.54)	(2.65–6.22)	(3.11–12.07)	(0.53–10.84)
A foreign national?						
Yes	3.55%	5.39%	2.96%	1.80%	1.78%	NR
	(2.14–5.51)	(2.73–9.43)	(0.85–7.24)	(0.80–3.46)	(0.39–4.98)	(2.44–56.36)
No	91.04%	88.78%	93.02%	92.31%	90.57%	NR
	(87.54–93.83)	(80.96–94.19)	(88.07–96.37)	(89.13–94.80)	(77.53–97.39)	(42.68–93.25)
Do not know	5.41%	5.83%	4.02%	5.89%	7.65%	6.41%
	(3.10–8.67)	(1.58–14.41)	(1.82–7.59)	(3.69–8.84)	(1.49–21.35)	(1.87–15.20)

NOTE: Includes estimates for active-component DoD and Coast Guard members. 95-percent confidence intervals for each estimate are included in parentheses.

NR = Not reportable.

Table A.12.c

Identity of offender(s) in the single or most serious assault among members who experienced a sexual assault in the past year, by pay grade

SAFU8g–k: At the time of the event, was the person/were the people who did this to you...

	Total	E1–E4	E5–E9	O1–O3	O4–O6
Someone in the military?					
Yes	NR	NR	NR	NR	NR
	(65.34–91.23)	(71.76–96.42)	(71.75–97.96)	(NR)	(NR)
No	16.40%	6.54%	NR	NR	NR
	(6.71–31.24)	(1.31–18.27)	(2.04–28.25)	(NR)	(NR)
Do not know	3.00%	NR	NR	NR	NR
	(0.38–10.26)	(0.35–23.55)	(0.00–38.25)	(NR)	(NR)
Civilian employee(s) or contractor(s) working for the military?					
Yes	NR	0.00%	NR	NR	NR
	(1.05–55.32)	(0.00–10.98)	(3.27–78.49)	(NR)	(NR)
No	NR	NR	NR	NR	NR
	(45.23–96.56)	(72.95–99.23)	(21.52–96.42)	(NR)	(NR)
Do not know	4.78%	NR	0.58%	NR	NR
	(1.16–12.50)	(0.77–27.05)	(0.00–13.85)	(NR)	(NR)
Person(s) in the local community?					
Yes	NR	12.02%	NR	NR	NR
	(12.29–59.31)	(3.76–26.69)	(14.40–82.41)	(NR)	(NR)
No	NR	NR	NR	NR	NR
	(38.77–84.52)	(63.14–93.08)	(17.40–85.12)	(NR)	(NR)
Do not know	3.41%	NR	0.54%	NR	NR
	(0.53–10.84)	(0.42–26.35)	(0.00–13.76)	(NR)	(NR)
A foreign national?					
Yes	NR	0.00%	NR	NR	NR
	(2.44–56.36)	(0.00–10.98)	(1.90–78.73)	(NR)	(NR)
No	NR	NR	NR	NR	NR
	(42.68–93.25)	(63.59–95.53)	(21.36–97.86)	(NR)	(NR)
Do not know	6.41%	NR	0.58%	NR	NR
	(1.87–15.20)	(4.47–36.41)	(0.00–13.85)	(NR)	(NR)

NOTE: Includes estimates for active-component Coast Guard members. 95-percent confidence intervals for each estimate are included in parentheses.

Too few warrant officers were included in the sample to break them out as a separate pay grade. For the purposes of this table, warrant officers have been included with the E5–E9 category.

NR = Not reportable.

A.13. Highest rank of offender(s) in the single or most serious assault among members who experienced a sexual assault in the past year and indicated the offender(s) included someone in the military

Table A.13.a

Highest rank of offender(s) in the single or most serious assault among members who experienced a sexual assault in the past year and indicated the offender(s) included someone in the military, by gender

Derived variable:

Responses scored so that the highest rank indicated on SAFU9a, SAFU9b, and SAFU9c ("You said a person/people in the military did this to you. Were any of them...") becomes the respondent's score, for mutually exclusive categories.

	Total	Men	Women
A lower rank than you	9.74% (3.54–20.35)	NR (NR)	11.17% (4.03–23.26)
A similar rank as you	NR (15.89–58.56)	NR (NR)	NR (17.96–48.38)
A higher rank than you	NR (30.82–76.28)	NR (NR)	NR (38.92–71.04)
Do not know	1.01% (0.01–7.22)	NR (NR)	1.68% (0.04–9.06)

NOTE: Includes estimates for active-component Coast Guard members. 95-percent confidence intervals for each estimate are included in parentheses.

NR = Not reportable.

Table A.13.b
Highest rank of offender(s) in the single or most serious assault among members who experienced a sexual assault in the past year and indicated the offender(s) included someone in the military, by service

Derived variable:

Responses scored so that the highest rank indicated on SAFU9a, SAFU9b, and SAFU9c ("You said a person/people in the military did this to you. Were any of them…") becomes the respondent's score, for mutually exclusive categories.

	Total DoD	Army	Navy	Air Force	Marine Corps	Coast Guard
A lower rank than you	8.77% (6.33–11.77)	8.92% (5.86–12.88)	10.27% (5.26–17.59)	9.62% (6.86–13.02)	3.43% (1.30–7.22)	9.74% (3.54–20.35)
A similar rank as you	34.83% (29.80–40.13)	30.68% (23.38–38.77)	33.74% (25.33–42.99)	42.93% (37.05–48.96)	NR (24.85–60.71)	NR (15.89–58.56)
A higher rank than you	54.26% (48.47–59.96)	58.91% (50.73–66.74)	53.89% (42.65–64.83)	43.91% (37.96–49.98)	NR (34.13–69.12)	NR (30.82–76.28)
Do not know	2.14% (1.30–3.31)	1.49% (0.31–4.28)	2.10% (0.97–3.95)	3.55% (2.04–5.69)	2.77% (0.48–8.46)	1.01% (0.01–7.22)

NOTE: Includes estimates for active-component DoD and Coast Guard members. 95-percent confidence intervals for each estimate are included in parentheses.

NR = Not reportable.

Table A.13.c
Highest rank of offender(s) in the single or most serious assault among members who experienced a sexual assault in the past year and indicated the offender(s) included someone in the military, by pay grade

Derived variable:

Responses scored so that the highest rank indicated on SAFU9a, SAFU9b, and SAFU9c ("You said a person/people in the military did this to you. Were any of them…") becomes the respondent's score, for mutually exclusive categories.

	Total	E1–E4	E5–E9	O1–O3	O4–O6
A lower rank than you	9.74% (3.54–20.35)	NR (0.98–23.85)	NR (1.01–29.68)	NR (NR)	NR (NR)
A similar rank as you	NR (15.89–58.56)	NR (20.12–57.89)	NR (7.79–73.21)	NR (NR)	NR (NR)
A higher rank than you	NR (30.82–76.28)	NR (34.48–72.65)	NR (18.65–87.66)	NR (NR)	NR (NR)
Do not know	1.01% (0.01–7.22)	0.93% (0.00–11.46)	NR (0.00–17.28)	NR (NR)	NR (NR)

NOTE: Includes estimates for active-component Coast Guard members. 95-percent confidence intervals for each estimate are included in parentheses.

Too few warrant officers were included in the sample to break them out as a separate pay grade. For the purposes of this table, warrant officers have been included with the E5–E9 category.

NR = Not reportable.

A.14. Percentage of members who indicated that the offender was an officer in the single or most serious assault among members who experienced a sexual assault in the past year and indicated that the offender(s) included someone in the military

Table A.14.a
Percentage of members who indicated that the offender was an officer in the single or most serious assault among members who experienced a sexual assault in the past year and indicated that the offender(s) included someone in the military, by gender

SAFU9d. You said a person/people in the military did this to you. Were any of them officers?

	Total	Men	Women
Yes	NR (11.74–68.83)	NR (NR)	NR (11.24–45.59)
No	NR (30.79–87.14)	NR (NR)	NR (53.04–87.15)
Do not know	1.16% (0.01–8.00)	NR (NR)	1.82% (0.04–9.83)

NOTE: Includes estimates for active-component Coast Guard members. 95-percent confidence intervals for each estimate are included in parentheses.

NR = Not reportable.

Table A.14.b
Percentage of members who indicated that the offender was an officer in the single or most serious assault among members who experienced a sexual assault in the past year and indicated that the offender(s) included someone in the military, by service

SAFU9d. You said a person/people in the military did this to you. Were any of them officers?

	Total DoD	Army	Navy	Air Force	Marine Corps	Coast Guard
Yes	15.22% (10.24–21.42)	16.78% (12.51–21.80)	NR (4.61–31.65)	20.48% (15.50–26.24)	8.72% (4.32–15.31)	NR (11.74–68.83)
No	81.42% (74.96–86.81)	79.13% (71.50–85.50)	NR (67.28–93.78)	77.59% (71.75–82.73)	84.28% (69.62–93.70)	NR (30.79–87.14)
Do not know	3.36% (1.29–7.04)	4.10% (0.42–14.85)	1.93% (0.78–3.93)	1.92% (0.74–4.05)	NR (0.65–25.05)	1.16% (0.01–8.00)

NOTE: Includes estimates for active-component DoD and Coast Guard members. 95-percent confidence intervals for each estimate are included in parentheses.

NR = Not reportable.

Table A.14.c
Percentage of members who indicated that the offender was an officer in the single or most serious assault among members who experienced a sexual assault in the past year and indicated that the offender(s) included someone in the military, by pay grade

SAFU9d. You said a person/people in the military did this to you. Were any of them officers?

	Total	E1–E4	E5–E9	O1–O3	O4–O6
Yes	NR	NR	NR	NR	NR
	(11.74–68.83)	(0.12–19.76)	(19.18–89.67)	(NR)	(NR)
No	NR	94.95%	NR	NR	NR
	(30.79–87.14)	(80.47–99.61)	(10.10–79.99)	(NR)	(NR)
Do not know	1.16%	1.04%	NR	NR	NR
	(0.01–8.00)	(0.00–12.78)	(0.00–19.04)	(NR)	(NR)

NOTE: Includes estimates for active-component Coast Guard members. 95-percent confidence intervals for each estimate are included in parentheses.

Too few warrant officers were included in the sample to break them out as a separate pay grade. For the purposes of this table, warrant officers have been included with the E5–E9 category.

NR = Not reportable.

A.15. Offender was respondents' unit leader or someone in their chain of command in the single or most serious assault among members who experienced a sexual assault in the past year and indicated that the offender was a higher rank

Table A.15.a
Offender was respondents' unit leader or someone in their chain of command in the single or most serious assault among members who experienced a sexual assault in the past year and indicated that the offender was a higher rank, by gender

SAFU9e. Was the higher ranked person your unit leader or someone above them in your chain of command?

	Total	Men	Women
Yes	NR (24.38–86.72)	NR (NR)	NR (16.25–58.60)
No	NR (13.28–75.62)	NR (NR)	NR (41.40–83.75)

NOTE: Includes estimates for active-component Coast Guard members. 95-percent confidence intervals for each estimate are included in parentheses.

Some response options that appear in Table A.15.b do not appear here because they were not selected by any Coast Guard respondent.

NR = Not reportable.

Table A.15.b
Offender was respondents' unit leader or someone in their chain of command in the single or most serious assault among members who experienced a sexual assault in the past year and indicated that the offender was a higher rank, by service

SAFU9e. Was the higher ranked person your unit leader or someone above them in your chain of command?

	Total DoD	Army	Navy	Air Force	Marine Corps	Coast Guard
Yes	32.57% (24.03–42.05)	30.46% (21.05–41.23)	NR (17.86–57.35)	26.58% (18.86–35.52)	NR (17.67–51.80)	NR (24.38–86.72)
No	61.66% (52.19–70.52)	64.80% (53.96–74.63)	NR (40.30–79.64)	72.28% (63.33–80.08)	NR (24.54–69.01)	NR (13.28–75.62)
Do not know	5.77% (2.89–10.16)	4.74% (2.20–8.79)	2.83% (0.93–6.46)	1.15% (0.17–3.82)	NR (5.17–47.09)	NR (NR)

NOTE: Includes estimates for active-component DoD and Coast Guard members. 95-percent confidence intervals for each estimate are included in parentheses.

NR = Not reportable.

Table A.15.c
Offender was respondents' unit leader or someone in their chain of command in the single or most serious assault among members who experienced a sexual assault in the past year and indicated that the offender was a higher rank, by pay grade

SAFU9e. Was the higher ranked person your unit leader or someone above them in your chain of command?

	Total	E1–E4	E5–E9	O1–O3	O4–O6
Yes	NR (24.38–86.72)	NR (14.74–64.42)	NR (NR)	NR (NR)	NR (NR)
No	NR (13.28–75.62)	NR (35.58–85.26)	NR (NR)	NR (NR)	NR (NR)

NOTE: Includes estimates for active-component Coast Guard members. 95-percent confidence intervals for each estimate are included in parentheses.

Too few warrant officers were included in the sample to break them out as a separate pay grade. For the purposes of this table, warrant officers have been included with the E5–E9 category.

Some response options that appear in Table A.15.b do not appear here because they were not selected by any Coast Guard respondent.

NR = Not reportable.

A.16. Location of the single or most serious assault among members who experienced a sexual assault in the past year

Table A.16.a
Location of the single or most serious assault among members who experienced a sexual assault in the past year, by gender

SAFU10: Did the unwanted event occur...

	Total	Men	Women
At a military installation/ship?	NR (35.11–72.96)	NR (NR)	NR (22.20–50.65)
During your work day/duty hours?	NR (24.21–67.33)	NR (NR)	18.32% (9.35–30.74)
While you were on TDY/TAD, at sea, or during field exercises/alerts?	NR (11.90–57.04)	NR (NR)	27.08% (16.12–40.56)
While you were deployed to a combat zone or to an area where you drew imminent danger pay or hostile fire pay?	NR (4.03–55.30)	NR (NR)	4.55% (0.56–15.40)
While you were in a delayed entry program?	NR (0.54–55.03)	NR (NR)	0.00% (0.00–5.74)
While you were in recruit training/basic training?	NR (0.55–55.34)	NR (NR)	0.00% (0.00–5.84)
While you were in any type of military combat training?	NR (0.54–55.03)	NR (NR)	0.00% (0.00–5.74)
While you were in Officer Candidate or Training School/Basic or Advanced Officer Course?	NR (1.07–54.10)	NR (NR)	2.72% (0.30–9.79)
While you were completing military occupational specialty school/technical training/advanced individual training/professional military education?	NR (2.89–53.56)	NR (NR)	8.77% (2.81–19.61)

NOTE: Includes estimates for active-component Coast Guard members. 95-percent confidence intervals for each estimate are included in parentheses.

NR = Not reportable.

Table A.16.b
Location of the single or most serious assault among members who experienced a sexual assault in the past year, by service

SAFU10: Did the unwanted event occur...

	Total DoD	Army	Navy	Air Force	Marine Corps	Coast Guard
At a military installation/ship?	65.29% (60.05–70.27)	71.01% (63.83–77.49)	65.60% (55.29–74.93)	50.76% (44.98–56.52)	NR (44.15–79.34)	NR (35.11–72.96)
During your work day/duty hours?	48.76% (43.26–54.28)	54.25% (46.54–61.82)	52.11% (41.46–62.61)	30.31% (24.73–36.36)	NR (25.88–59.90)	NR (24.21–67.33)
While you were on TDY/TAD, at sea, or during field exercises/alerts?	19.11% (14.38–24.59)	15.32% (10.25–21.64)	23.25% (13.68–35.37)	12.78% (9.20–17.13)	NR (10.21–42.26)	NR (11.90–57.04)
While you were deployed to a combat zone or to an area where you drew imminent danger pay or hostile fire pay?	14.67% (10.80–19.29)	11.75% (8.29–16.02)	12.69% (8.45–18.05)	10.53% (7.09–14.89)	NR (12.96–55.23)	NR (4.03–55.30)
While you were in a delayed entry program?	2.22% (1.09–4.01)	3.10% (1.36–5.96)	0.61% (0.12–1.82)	0.56% (0.13–1.56)	5.83% (0.81–18.72)	NR (0.54–55.03)
While you were in recruit training/basic training?	5.07% (2.66–8.64)	8.25% (3.97–14.79)	0.54% (0.10–1.62)	1.35% (0.37–3.42)	NR (2.41–33.37)	NR (0.55–55.34)
While you were in any type of military combat training?	5.72% (3.69–8.42)	9.17% (5.39–14.35)	1.56% (0.61–3.28)	3.32% (0.77–8.95)	NR (2.47–25.81)	NR (0.54–55.03)
While you were in Officer Candidate or Training School/Basic or Advanced Officer Course?	3.51% (1.24–7.66)	3.23% (1.54–5.90)	0.53% (0.05–2.07)	0.56% (0.13–1.55)	NR (2.36–41.39)	NR (1.07–54.10)
While you were completing military occupational specialty school/technical training/advanced individual training/professional military education?	11.97% (8.39–16.38)	11.39% (6.72–17.71)	7.68% (4.76–11.61)	11.51% (8.18–15.58)	NR (8.96–49.31)	NR (2.89–53.56)

NOTE: Includes estimates for active-component DoD and Coast Guard members. 95-percent confidence intervals for each estimate are included in parentheses.

NR = Not reportable.

Table A.16.c

Location of the single or most serious assault among members who experienced a sexual assault in the past year, by pay grade

SAFU10: Did the unwanted event occur...

	Total	E1–E4	E5–E9	O1–O3	O4–O6
At a military installation/ship?	NR (35.11–72.96)	NR (19.67–55.13)	NR (63.40–95.15)	NR (NR)	NR (NR)
During your work day/duty hours?	NR (24.21–67.33)	NR (6.96–37.26)	NR (37.03–89.07)	NR (NR)	NR (NR)
While you were on TDY/TAD, at sea, or during field exercises/alerts?	NR (11.90–57.04)	NR (18.57–53.58)	NR (4.24–78.47)	NR (NR)	NR (NR)
While you were deployed to a combat zone or to an area where you drew imminent danger pay or hostile fire pay?	NR (4.03–55.30)	0.00% (0.00–9.75)	NR (3.25–78.52)	NR (NR)	NR (NR)
While you were in a delayed entry program?	NR (0.54–55.03)	0.00% (0.00–9.75)	NR (1.91–78.68)	NR (NR)	NR (NR)
While you were in recruit training/basic training?	NR (0.55–55.34)	0.00% (0.00–9.75)	NR (1.91–78.69)	NR (NR)	NR (NR)
While you were in any type of military combat training?	NR (0.54–55.03)	0.00% (0.00–9.75)	NR (1.91–78.68)	NR (NR)	NR (NR)
While you were in Officer Candidate or Training School/Basic or Advanced Officer Course?	NR (1.07–54.10)	3.45% (0.10–17.26)	NR (2.14–78.59)	NR (NR)	NR (NR)
While you were completing military occupational specialty school/technical training/advanced individual training/professional military education?	NR (2.89–53.56)	NR (1.22–27.82)	NR (2.74–78.47)	NR (NR)	NR (NR)

NOTE: Includes estimates for active-component Coast Guard members. 95-percent confidence intervals for each estimate are included in parentheses.

Too few warrant officers were included in the sample to break them out as a separate pay grade. For the purposes of this table, warrant officers have been included with the E5–E9 category.

NR = Not reportable.

A.17. Situational context of the single or most serious assault among members who experienced a sexual assault in the past year

Table A.17.a

Situational context of the single or most serious assault among members who experienced a sexual assault in the past year, by gender

SAFU11: Which of the following best describes the situation when this unwanted event occurred?

	Total	Men	Women
You were out with friends or at a party.	NR (31.10–70.45)	NR (NR)	50.79% (36.92–64.57)
You were on a date.	NR (1.19–52.99)	NR (NR)	3.23% (0.55–9.93)
You were being intimate with the other person.	NR (1.04–53.09)	NR (NR)	2.62% (0.33–8.97)
You were at work.	NR (24.26–66.64)	NR (NR)	17.25% (8.72–29.21)
You were alone in a public place.	NR (4.22–52.90)	NR (NR)	12.00% (4.73–23.79)
You were in your home or quarters.	NR (6.84–53.89)	NR (NR)	15.48% (7.31–27.40)
You were in someone else's home or quarters.	NR (7.62–54.12)	NR (NR)	19.00% (10.29–30.72)
You were at a military function.	NR (1.24–53.03)	NR (NR)	3.46% (0.36–12.59)
You were in temporary lodging/hotel.	NR (4.47–52.81)	NR (NR)	12.47% (5.61–22.96)
None of the above.	3.45% (0.67–10.02)	NR (NR)	3.00% (0.47–9.54)

NOTE: Includes estimates for active-component Coast Guard members. 95-percent confidence intervals for each estimate are included in parentheses.

Some response options that appear in Table A.17.b do not appear here because they were not selected by any Coast Guard respondent.

NR = Not reportable.

Table A.17.b
Situational context of the single or most serious assault among members who experienced a sexual assault in the past year, by service

SAFU11: Which of the following best describes the situation when this unwanted event occurred?

	Total DoD	Army	Navy	Air Force	Marine Corps	Coast Guard
You were out with friends or at a party.	29.63% (25.14–34.43)	25.17% (19.49–31.56)	35.51% (25.88–46.08)	39.21% (33.73–44.90)	18.12% (11.20–26.97)	NR (31.10–70.45)
You were on a date.	3.07% (1.98–4.54)	2.52% (1.50–3.97)	3.47% (1.19–7.71)	2.78% (1.62–4.43)	3.70% (1.45–7.65)	NR (1.19–52.99)
You were being intimate with the other person.	5.06% (3.36–7.27)	4.06% (1.47–8.74)	6.80% (3.51–11.68)	5.07% (3.22–7.55)	3.24% (1.28–6.66)	NR (1.04–53.09)
You were at work.	43.22% (37.84–48.71)	45.18% (37.30–53.25)	47.53% (37.33–57.89)	24.71% (19.35–30.72)	NR (26.61–61.55)	NR (24.26–66.64)
You were alone in a public place.	10.76% (8.44–13.46)	12.52% (8.02–18.34)	9.39% (6.08–13.70)	11.72% (8.28–15.95)	8.96% (4.40–15.82)	NR (4.22–52.90)
You were in your home or quarters.	17.11% (14.25–20.27)	17.52% (12.17–24.02)	16.73% (12.13–22.23)	21.67% (18.11–25.59)	13.13% (7.87–20.13)	NR (6.84–53.89)
You were in someone else's home or quarters.	17.73% (14.80–20.98)	16.31% (12.63–20.57)	21.08% (14.63–28.80)	20.01% (16.30–24.15)	11.12% (6.55–17.31)	NR (7.62–54.12)
You were at a military function.	16.85% (11.83–22.92)	14.34% (8.67–21.82)	17.45% (8.13–30.94)	11.20% (7.52–15.86)	NR (10.76–47.86)	NR (1.24–53.03)
You were in temporary lodging/hotel.	5.25% (4.06–6.67)	4.37% (2.54–6.96)	5.28% (3.18–8.18)	9.16% (6.54–12.40)	4.05% (1.80–7.72)	NR (4.47–52.81)
None of the above.	7.82% (4.62–12.22)	9.34% (4.37–16.97)	3.29% (1.70–5.71)	3.82% (2.44–5.68)	NR (5.06–41.54)	3.45% (0.67–10.02)
Do not recall.	2.76% (1.56–4.48)	3.49% (1.48–6.86)	2.10% (0.39–6.32)	1.92% (0.68–4.22)	3.26% (0.91–8.08)	0.00% (0.00–11.23)

NOTE: Includes estimates for active-component DoD and Coast Guard members. 95-percent confidence intervals for each estimate are included in parentheses.

NR = Not reportable.

Table A.17.c
Situational context of the single or most serious assault among members who experienced a sexual assault in the past year, by pay grade

SAFU11: Which of the following best describes the situation when this unwanted event occurred?

	Total	E1–E4	E5–E9	O1–O3	O4–O6
You were out with friends or at a party.	NR (31.10–70.45)	NR (40.09–74.56)	NR (20.46–84.81)	NR (NR)	NR (NR)
You were on a date.	NR (1.19–52.99)	0.79% (0.00–9.52)	NR (1.92–78.66)	NR (NR)	NR (NR)
You were being intimate with the other person.	NR (1.04–53.09)	1.84% (0.02–11.33)	NR (2.82–78.44)	NR (NR)	NR (NR)
You were at work.	NR (24.26–66.64)	NR (6.40–34.98)	NR (39.22–89.77)	NR (NR)	NR (NR)
You were alone in a public place.	NR (4.22–52.90)	NR (7.07–37.53)	NR (2.38–78.50)	NR (NR)	NR (NR)
You were in your home or quarters.	NR (6.84–53.89)	NR (4.13–31.11)	NR (5.59–78.79)	NR (NR)	NR (NR)
You were in someone else's home or quarters.	NR (7.62–54.12)	NR (14.05–43.49)	NR (3.13–78.49)	NR (NR)	NR (NR)
You were at a military function.	NR (1.24–53.03)	NR (0.40–21.04)	NR (1.92–78.66)	NR (NR)	NR (NR)
You were in temporary lodging/hotel.	NR (4.47–52.81)	NR (4.50–29.51)	NR (2.72–78.45)	NR (NR)	NR (NR)
None of the above.	3.45% (0.67–10.02)	0.00% (0.00–9.19)	NR (0.31–21.11)	NR (NR)	NR (NR)

NOTE: Includes estimates for active-component Coast Guard members. 95-percent confidence intervals for each estimate are included in parentheses.

Too few warrant officers were included in the sample to break them out as a separate pay grade. For the purposes of this table, warrant officers have been included with the E5–E9 category.

Some response options that appear in Table A.17.b do not appear here because they were not selected by any Coast Guard respondent.

NR = Not reportable.

A.18. Percentage of the single or most serious assaults considered to be hazing among members who experienced a sexual assault in the past year

Table A.18.a
Percentage of the single or most serious assaults considered to be hazing among members who experienced a sexual assault in the past year, by gender

SAFU12: Would you describe this unwanted event as hazing? Hazing refers to things done to humiliate or 'toughen up' people prior to accepting them into a group.

Total	Men	Women
NR	NR	2.70%
(1.47–52.92)	(NR)	(0.23–10.60)

NOTE: Includes estimates for active-component Coast Guard members. 95-percent confidence intervals for each estimate are included in parentheses.

NR = Not reportable.

Table A.18.b
Percentage of the single or most serious assaults considered to be hazing among members who experienced a sexual assault in the past year, by service

SAFU12: Would you describe this unwanted event as hazing? Hazing refers to things done to humiliate or 'toughen up' people prior to accepting them into a group.

Total DoD	Army	Navy	Air Force	Marine Corps	Coast Guard
19.82%	19.69%	NR	10.39%	NR	NR
(14.13–26.58)	(12.45–28.79)	(12.33–39.14)	(5.73–16.95)	(6.47–36.87)	(1.47–52.92)

NOTE: Includes estimates for active-component DoD and Coast Guard members. 95-percent confidence intervals for each estimate are included in parentheses.

NR = Not reportable.

Table A.18.c
Percentage of the single or most serious assaults considered to be hazing among members who experienced a sexual assault in the past year, by pay grade

SAFU12: Would you describe this unwanted event as hazing? Hazing refers to things done to humiliate or 'toughen up' people prior to accepting them into a group.

Total	E1–E4	E5–E9	O1–O3	O4–O6
NR (1.47–52.92)	3.55% (0.10–17.73)	NR (3.28–78.45)	NR (NR)	NR (NR)

NOTE: Includes estimates for active-component Coast Guard members. 95-percent confidence intervals for each estimate are included in parentheses.

Too few warrant officers were included in the sample to break them out as a separate pay grade. For the purposes of this table, warrant officers have been included with the E5–E9 category.

NR = Not reportable.

A.19. Stalking and/or harassment by the offender before or after the single or most serious assault among members who experienced a sexual assault in the past year

Table A.19.a
Stalking and/or harassment by the offender before or after the single or most serious assault among members who experienced a sexual assault in the past year, by gender

SAFU13: Did the offender(s)...

	Total	Men	Women
Sexually harass you before the situation?	NR (25.61–67.21)	NR (NR)	30.01% (18.37–43.89)
Stalk you before the situation?	NR (14.89–60.51)	NR (NR)	NR (12.64–41.38)
Sexually harass you after the situation?	NR (13.16–59.76)	NR (NR)	14.97% (7.07–26.53)
Stalk you after the situation?	NR (11.44–58.87)	NR (NR)	12.10% (5.08–23.17)

NOTE: Includes estimates for active-component Coast Guard members. 95-percent confidence intervals for each estimate are included in parentheses.

NR = Not reportable.

Table A.19.b
Stalking and/or harassment by the offender before or after the single or most serious assault among members who experienced a sexual assault in the past year, by service

SAFU13: Did the offender(s)...

	Total DoD	Army	Navy	Air Force	Marine Corps	Coast Guard
Sexually harass you before the situation?	34.23% (29.02–39.75)	35.23% (27.79–43.24)	33.95% (23.91–45.18)	31.27% (26.15–36.76)	NR (19.47–53.22)	NR (25.61–67.21)
Stalk you before the situation?	9.22% (7.09–11.72)	12.50% (9.00–16.74)	6.09% (3.09–10.61)	8.04% (5.67–10.99)	9.75% (3.22–21.36)	NR (14.89–60.51)
Sexually harass you after the situation?	34.10% (28.49–40.06)	35.79% (28.23–43.90)	36.43% (25.18–48.87)	25.84% (20.90–31.28)	NR (15.96–49.94)	NR (13.16–59.76)
Stalk you after the situation?	11.64% (9.37–14.24)	13.70% (10.12–17.96)	9.50% (6.02–14.08)	11.48% (8.54–14.98)	12.13% (4.98–23.52)	NR (11.44–58.87)

NOTE: Includes estimates for active-component DoD and Coast Guard members. 95-percent confidence intervals for each estimate are included in parentheses.

NR = Not reportable.

Table A.19.c
Stalking and/or harassment by the offender before or after the single or most serious assault among members who experienced a sexual assault in the past year, by pay grade

SAFU13: Did the offender(s)...

	Total	E1–E4	E5–E9	O1–O3	O4–O6
Sexually harass you before the situation?	NR (25.61–67.21)	NR (23.76–58.88)	NR (31.46–87.27)	NR (NR)	NR (NR)
Stalk you before the situation?	NR (14.89–60.51)	NR (11.45–45.41)	NR (10.49–81.45)	NR (NR)	NR (NR)
Sexually harass you after the situation?	NR (13.16–59.76)	NR (8.92–37.60)	NR (10.63–81.73)	NR (NR)	NR (NR)
Stalk you after the situation?	NR (11.44–58.87)	NR (5.20–32.32)	NR (10.79–81.76)	NR (NR)	NR (NR)

NOTE: Includes estimates for active-component Coast Guard members. 95-percent confidence intervals for each estimate are included in parentheses.

Too few warrant officers were included in the sample to break them out as a separate pay grade. For the purposes of this table, warrant officers have been included with the E5–E9 category.

NR = Not reportable.

A.20. Involvement of alcohol and/or drugs in the single or most serious assault among members who experienced a sexual assault in the past year

Table A.20.a
Involvement of alcohol and/or drugs in the single or most serious assault among members who experienced a sexual assault in the past year, by gender

SAFU14–SAFU17

	Total	Men	Women
At the time of this unwanted event had you been drinking alcohol? Even if you had been drinking, it does not mean that you are to blame for what happened.			
Yes	NR (35.62–74.10)	NR (NR)	63.35% (49.78–75.52)
No	NR (25.68–64.11)	NR (NR)	36.21% (24.08–49.77)
Not sure	0.28% (0.00–4.70)	NR (NR)	0.45% (0.00–5.48)
Just prior to this unwanted event, did the person(s) who did this to you buy or give you alcohol to drink?			
Yes	NR (44.93–87.23)	NR (NR)	NR (38.16–74.45)
No	NR (8.27–42.62)	NR (NR)	NR (15.62–49.21)
Do not know	9.00% (2.03–23.45)	NR (NR)	NR (3.53–28.81)
Just prior to this unwanted event, do you think that you might have been given a drug without your knowledge or consent?			
Yes	NR (2.71–52.62)	NR (NR)	8.16% (2.77–17.81)
No	NR (45.38–91.83)	NR (NR)	79.82% (67.10–89.29)
Do not know	7.59% (2.68–16.28)	NR (NR)	12.01% (4.78–23.70)
At the time of this unwanted event, had the person(s) who did it been drinking alcohol?			
Yes	NR (43.12–78.30)	NR (NR)	62.31% (48.33–74.94)
No	NR (12.28–40.43)	NR (NR)	23.29% (13.26–36.13)
Do not know	13.78% (4.85–28.65)	NR (NR)	14.40% (6.38–26.52)

NOTE: Includes estimates for active-component Coast Guard members.
95-percent confidence intervals for each estimate are included in parentheses.

NR = Not reportable.

Table A.20.b
Involvement of alcohol and/or drugs in the single or most serious assault among members who experienced a sexual assault in the past year, by service

SAFU14–SAFU17

	Total DoD	Army	Navy	Air Force	Marine Corps	Coast Guard
At the time of this unwanted event had you been drinking alcohol? Even if you had been drinking, it does not mean that you are to blame for what happened.						
Yes	33.49%	26.91%	34.29%	45.76%	NR	NR
	(28.81–38.42)	(21.67–32.67)	(25.01–44.53)	(40.04–51.56)	(22.08–54.80)	(35.62–74.10)
No	62.22%	65.56%	63.28%	52.80%	NR	NR
	(57.06–67.18)	(57.97–72.61)	(53.06–72.69)	(46.99–58.56)	(42.16–75.06)	(25.68–64.11)
Not sure	4.29%	7.53%	2.44%	1.44%	3.23%	0.28%
	(2.08–7.75)	(2.39–17.03)	(0.85–5.40)	(0.60–2.88)	(1.22–6.83)	(0.00–4.70)
Just prior to this unwanted event, did the person(s) who did this to you buy or give you alcohol to drink?						
Yes	51.33%	58.85%	NR	51.34%	NR	NR
	(42.74–59.86)	(48.93–68.27)	(24.38–52.91)	(43.16–59.47)	(46.42–85.15)	(44.93–87.23)
No	36.24%	37.99%	NR	40.57%	NR	NR
	(29.38–43.53)	(28.67–48.01)	(23.30–52.70)	(32.79–48.71)	(11.89–46.45)	(8.27–42.62)
Do not know	12.43%	3.16%	NR	8.09%	5.36%	9.00%
	(4.36–26.10)	(1.37–6.14)	(6.40–54.56)	(4.90–12.42)	(1.60–12.67)	(2.03–23.45)
Just prior to this unwanted event, do you think that you might have been given a drug without your knowledge or consent?						
Yes	6.18%	7.68%	2.32%	2.80%	NR	NR
	(3.30–10.40)	(3.50–14.28)	(1.08–4.33)	(1.50–4.73)	(2.63–39.53)	(2.71–52.62)
No	85.52%	80.90%	92.72%	88.47%	NR	NR
	(81.16–89.19)	(72.85–87.41)	(89.52–95.18)	(85.24–91.21)	(56.50–90.88)	(45.38–91.83)
Do not know	8.30%	11.42%	4.96%	8.74%	8.42%	7.59%
	(6.07–11.04)	(6.40–18.38)	(3.05–7.57)	(6.40–11.59)	(4.34–14.45)	(2.68–16.28)
At the time of this unwanted event, had the person(s) who did it been drinking alcohol?						
Yes	37.36%	32.31%	39.12%	48.90%	NR	NR
	(32.56–42.34)	(26.26–38.83)	(29.64–49.24)	(43.10–54.72)	(20.84–52.73)	(43.12–78.30)
No	46.38%	49.66%	43.62%	36.02%	NR	NR
	(41.06–51.77)	(41.83–57.50)	(33.48–54.17)	(30.33–42.01)	(36.84–70.28)	(12.28–40.43)
Do not know	16.26%	18.03%	17.26%	15.08%	10.50%	13.78%
	(12.64–20.43)	(11.75–25.87)	(10.83–25.45)	(11.04–19.90)	(4.02–21.26)	(4.85–28.65)

NOTE: Includes estimates for active-component DoD and Coast Guard members. 95-percent confidence intervals for each estimate are included in parentheses.

NR = Not reportable.

Table A.20.c
Involvement of alcohol and/or drugs in the single or most serious assault among members who experienced a sexual assault in the past year, by pay grade

SAFU14–SAFU17

	Total	E1–E4	E5–E9	O1–O3	O4–O6
At the time of this unwanted event had you been drinking alcohol? Even if you had been drinking, it does not mean that you are to blame for what happened.					
Yes	NR	NR	NR	NR	NR
	(35.62–74.10)	(51.93–84.24)	(19.55–84.70)	(NR)	(NR)
No	NR	NR	NR	NR	NR
	(25.68–64.11)	(15.09–47.35)	(15.30–80.45)	(NR)	(NR)
Not sure	0.28%	0.78%	NR	NR	NR
	(0.00–4.70)	(0.00–9.51)	(0.00–38.25)	(NR)	(NR)
Just prior to this unwanted event, did the person(s) who did this to you buy or give you alcohol to drink?					
Yes	NR	NR	NR	NR	NR
	(44.93–87.23)	(28.02–70.07)	(NR)	(NR)	(NR)
No	NR	NR	NR	NR	NR
	(8.27–42.62)	(14.35–54.99)	(NR)	(NR)	(NR)
Do not know	9.00%	NR	NR	NR	NR
	(2.03–23.45)	(5.36–41.65)	(NR)	(NR)	(NR)
Just prior to this unwanted event, do you think that you might have been given a drug without your knowledge or consent?					
Yes	NR	8.54%	NR	NR	NR
	(2.71–52.62)	(1.76–23.19)	(3.06–78.45)	(NR)	(NR)
No	NR	NR	NR	NR	NR
	(45.38–91.83)	(63.09–92.26)	(20.82–92.94)	(NR)	(NR)
Do not know	7.59%	NR	NR	NR	NR
	(2.68–16.28)	(2.62–27.13)	(0.67–23.19)	(NR)	(NR)
At the time of this unwanted event, had the person(s) who did it been drinking alcohol?					
Yes	NR	NR	NR	NR	NR
	(43.12–78.30)	(51.01–84.64)	(33.53–88.61)	(NR)	(NR)
No	NR	NR	NR	NR	NR
	(12.28–40.43)	(6.43–35.10)	(6.71–43.15)	(NR)	(NR)
Do not know	13.78%	NR	NR	NR	NR
	(4.85–28.65)	(3.45–29.88)	(1.76–43.77)	(NR)	(NR)

NOTE: Includes estimates for active-component Coast Guard members. 95-percent confidence intervals for each estimate are included in parentheses.

Too few warrant officers were included in the sample to break them out as a separate pay grade. For the purposes of this table, warrant officers have been included with the E5–E9 category.

NR = Not reportable.

A.21. Consequences of the single or most serious assault among members who experienced a sexual assault in the past year

Table A.21.a
Consequences of the single or most serious assault among members who experienced a sexual assault in the past year, by gender

SAFU18: After this unwanted event:

	Total	Men	Women
Did you request a transfer or other change of assignment as a result of the event?	NR (2.00–52.69)	NR (NR)	6.10% (1.66–15.02)
Did it make you want to leave the military?	NR (22.74–65.38)	NR (NR)	NR (20.22–48.27)
Did it ever make it hard to do your job or complete your work?	NR (31.57–70.78)	NR (NR)	46.13% (32.55–60.14)
Did you take a sick day or any other type of leave because of the event?	NR (9.37–55.39)	NR (NR)	22.50% (12.16–36.09)
Did the event damage your personal relationships, for example with your spouse or a friend?	NR (32.07–71.22)	NR (NR)	50.70% (36.78–64.54)

NOTE: Includes estimates for active-component Coast Guard members. 95-percent confidence intervals for each estimate are included in parentheses.

NR = Not reportable.

Table A.21.b
Consequences of the single or most serious assault among members who experienced a sexual assault in the past year, by service

SAFU18: After this unwanted event:

	Total DoD	Army	Navy	Air Force	Marine Corps	Coast Guard
Did you request a transfer or other change of assignment as a result of the event?	12.65% (9.23–16.77)	15.03% (10.21–21.01)	8.01% (4.20–13.61)	11.68% (8.31–15.80)	NR (5.55–41.89)	NR (2.00–52.69)
Did it make you want to leave the military?	35.11% (30.05–40.44)	41.21% (33.84–48.89)	33.87% (23.78–45.16)	25.43% (21.03–30.24)	NR (17.47–47.84)	NR (22.74–65.38)
Did it ever make it hard to do your job or complete your work?	45.50% (40.22–50.85)	48.05% (40.31–55.87)	46.24% (36.06–56.66)	42.26% (36.70–47.96)	NR (22.79–58.76)	NR (31.57–70.78)
Did you take a sick day or any other type of leave because of the event?	19.62% (14.65–25.41)	21.83% (15.86–28.81)	16.17% (7.11–29.70)	17.23% (13.46–21.56)	NR (9.04–48.27)	NR (9.37–55.39)
Did the event damage your personal relationships, for example with your spouse or a friend?	40.63% (35.34–46.08)	38.16% (31.11–45.59)	36.65% (26.78–47.42)	41.76% (36.43–47.23)	NR (39.41–71.69)	NR (32.07–71.22)

NOTE: Includes estimates for active-component DoD and Coast Guard members. 95-percent confidence intervals for each estimate are included in parentheses.

NR = Not reportable.

Table A.21.c
Consequences of the single or most serious assault among members who experienced a sexual assault in the past year, by pay grade

SAFU18: After this unwanted event:

	Total	E1–E4	E5–E9	O1–O3	O4–O6
Did you request a transfer or other change of assignment as a result of the event?	NR (2.00–52.69)	7.60% (1.63–20.47)	NR (2.93–78.48)	NR (NR)	NR (NR)
Did it make you want to leave the military?	NR (22.74–65.38)	NR (22.38–56.76)	NR (24.33–85.54)	NR (NR)	NR (NR)
Did it ever make it hard to do your job or complete your work?	NR (31.57–70.78)	NR (32.15–67.33)	NR (26.85–86.84)	NR (NR)	NR (NR)
Did you take a sick day or any other type of leave because of the event?	NR (9.37–55.39)	NR (11.16–43.85)	NR (5.82–78.86)	NR (NR)	NR (NR)
Did the event damage your personal relationships, for example with your spouse or a friend?	NR (32.07–71.22)	NR (24.66–59.40)	NR (27.55–88.06)	NR (NR)	NR (NR)

NOTE: Includes estimates for active-component Coast Guard members. 95-percent confidence intervals for each estimate are included in parentheses.

Too few warrant officers were included in the sample to break them out as a separate pay grade. For the purposes of this table, warrant officers have been included with the E5–E9 category.

NR = Not reportable.

A.22. Percentage of members who told someone about the single or most serious assault among members who experienced a sexual assault in the past year

Table A.22.a
Percentage of members who told someone about the single or most serious assault among members who experienced a sexual assault in the past year, by gender

SAFU19: Did you tell anyone about the unwanted event?

Total	Men	Women
NR	NR	66.28%
(46.06–81.63)	(NR)	(52.67–78.17)

NOTE: Includes estimates for active-component Coast Guard members. 95-percent confidence intervals for each estimate are included in parentheses.

NR = Not reportable.

Table A.22.b
Percentage of members who told someone about the single or most serious assault among members who experienced a sexual assault in the past year, by service

SAFU19: Did you tell anyone about the unwanted event?

Total DoD	Army	Navy	Air Force	Marine Corps	Coast Guard
50.84%	53.17%	47.24%	56.80%	NR	NR
(45.48–56.19)	(45.02–61.19)	(37.40–57.24)	(50.67–62.77)	(31.96–65.96)	(46.06–81.63)

NOTE: Includes estimates for active-component DoD and Coast Guard members. 95-percent confidence intervals for each estimate are included in parentheses.

NR = Not reportable.

Table A.22.c
Percentage of members who told someone about the single or most serious assault among members who experienced a sexual assault in the past year, by pay grade

SAFU19: Did you tell anyone about the unwanted event?

Total	E1–E4	E5–E9	O1–O3	O4–O6
NR	NR	NR	NR	NR
(46.06–81.63)	(40.74–75.31)	(30.67–91.28)	(NR)	(NR)

NOTE: Includes estimates for active-component Coast Guard members. 95-percent confidence intervals for each estimate are included in parentheses.

Too few warrant officers were included in the sample to break them out as a separate pay grade. For the purposes of this table, warrant officers have been included with the E5–E9 category.

NR = Not reportable.

A.23. Person(s) whom members told about the single or most serious assault among members who experienced a sexual assault in the past year

Table A.23.a
Person(s) whom members told about the single or most serious assault among members who experienced a sexual assault in the past year, by gender

SAFU20: Who have you talked to about this event?

	Total	Men	Women
A friend or family member	NR (35.31–74.31)	NR (NR)	59.10% (45.05–72.14)
The Sexual Assault Response Coordinator (SARC)	NR (6.96–54.41)	NR (NR)	15.51% (7.63–26.81)
A Sexual Assault Prevention and Response Victim Advocate (SAPR VA or VA)	NR (8.95–55.38)	NR (NR)	19.31% (10.27–31.51)
A Safe Helpline or other hotline counselor	NR (1.62–53.35)	NR (NR)	4.79% (0.97–13.50)
A medical professional (for example, a doctor or nurse)	NR (4.12–53.32)	NR (NR)	9.50% (3.74–19.04)
A chaplain or religious leader	NR (5.16–53.83)	NR (NR)	12.39% (4.80–24.74)
A counselor, therapist, or psychologist	NR (8.69–55.36)	NR (NR)	20.97% (11.32–33.78)
A Special Victims' Counsel or Victims' Legal Counsel	NR (2.70–53.15)	NR (NR)	5.96% (1.62–14.68)
Some other military lawyer (for example, a JAG officer)	NR (1.47–53.51)	NR (NR)	4.21% (0.87–11.85)
A supervisor or someone above you in your chain of command	NR (12.11–57.48)	NR (NR)	NR (12.15–39.45)
An officer or NCO outside of your chain of command	NR (5.30–55.31)	NR (NR)	5.20% (1.24–13.65)
Someone in military law enforcement	NR (2.28–53.24)	NR (NR)	6.85% (2.11–15.81)
Civilian law enforcement authority	NR (1.13–53.78)	NR (NR)	2.92% (0.35–10.22)
Civilian rape crisis center or other sexual assault advocacy group	NR (0.87–54.07)	NR (NR)	1.79% (0.10–7.89)

NOTE: Includes estimates for active-component Coast Guard members.
95-percent confidence intervals for each estimate are included in parentheses.

NR = Not reportable.

Table A.23.b
Person(s) whom members told about the single or most serious assault among members who experienced a sexual assault in the past year, by service

SAFU20: Who have you talked to about this event?

	Total DoD	Army	Navy	Air Force	Marine Corps	Coast Guard
A friend or family member	42.63% (37.76–47.61)	44.12% (36.68–51.76)	40.95% (31.93–50.44)	52.12% (46.15–58.05)	35.10% (22.09–49.97)	NR (35.31–74.31)
The Sexual Assault Response Coordinator (SARC)	14.78% (11.47–18.62)	17.92% (13.78–22.70)	8.78% (5.09–13.88)	15.82% (12.73–19.33)	NR (7.06–43.01)	NR (6.96–54.41)
A Sexual Assault Prevention and Response Victim Advocate (SAPR VA or VA)	13.17% (10.93–15.68)	16.26% (12.48–20.66)	10.23% (7.07–14.18)	13.14% (10.10–16.69)	13.04% (5.95–23.75)	NR (8.95–55.38)
A Safe Helpline or other hotline counselor	2.75% (1.51–4.58)	4.39% (2.36–7.39)	0.59% (0.12–1.74)	1.17% (0.45–2.44)	5.46% (0.66–18.39)	NR (1.62–53.35)
A medical professional (for example, a doctor or nurse)	11.35% (8.27–15.07)	12.79% (9.12–17.27)	7.58% (5.05–10.83)	8.13% (5.79–11.02)	NR (6.04–42.38)	NR (4.12–53.32)
A chaplain or religious leader	8.57% (6.46–11.10)	10.45% (7.43–14.17)	7.01% (3.59–12.12)	5.70% (3.41–8.88)	10.31% (3.74–21.52)	NR (5.16–53.83)
A counselor, therapist, or psychologist	12.89% (10.65–15.40)	17.16% (13.28–21.63)	8.12% (5.21–11.94)	12.27% (9.47–15.54)	14.93% (7.39–25.78)	NR (8.69–55.36)
A Special Victims' Counsel or Victims' Legal Counsel	8.94% (6.06–12.60)	9.39% (6.45–13.09)	4.45% (2.73–6.80)	9.49% (7.03–12.46)	NR (4.94–41.76)	NR (2.70–53.15)
Some other military lawyer (for example, a JAG officer)	6.18% (4.50–8.24)	8.57% (5.77–12.16)	4.03% (2.10–6.92)	5.47% (3.50–8.08)	6.37% (1.16–18.47)	NR (1.47–53.51)
A supervisor or someone above you in your chain of command	21.22% (17.21–25.69)	23.48% (18.31–29.31)	17.41% (11.13–25.34)	19.33% (15.30–23.89)	NR (11.64–47.11)	NR (12.11–57.48)
An officer or NCO outside of your chain of command	10.38% (8.12–13.02)	16.81% (12.51–21.88)	3.07% (1.42–5.75)	10.00% (6.84–13.98)	13.16% (5.09–26.20)	NR (5.30–55.31)
Someone in military law enforcement	7.20% (5.41–9.35)	8.88% (6.02–12.52)	5.97% (3.53–9.36)	7.25% (4.90–10.27)	6.16% (1.05–18.39)	NR (2.28–53.24)
Civilian law enforcement authority	3.41% (2.10–5.22)	5.19% (3.02–8.23)	1.25% (0.47–2.68)	2.80% (1.27–5.30)	4.97% (0.46–18.37)	NR (1.13–53.78)
Civilian rape crisis center or other sexual assault advocacy group	2.59% (1.36–4.44)	3.87% (1.94–6.85)	1.22% (0.45–2.64)	0.52% (0.10–1.54)	4.57% (0.30–18.58)	NR (0.87–54.07)

NOTE: Includes estimates for active-component DoD and Coast Guard members. 95-percent confidence intervals for each estimate are included in parentheses.

NR = Not reportable.

Table A.23.c

Person(s) whom members told about the single or most serious assault among members who experienced a sexual assault in the past year, by pay grade

SAFU20: Who have you talked to about this event?

	Total	E1–E4	E5–E9	O1–O3	O4–O6
A friend or family member	NR (35.31–74.31)	NR (29.25–64.26)	NR (29.39–91.17)	NR (NR)	NR (NR)
The Sexual Assault Response Coordinator (SARC)	NR (6.96–54.41)	NR (6.94–33.59)	NR (6.61–79.80)	NR (NR)	NR (NR)
A Sexual Assault Prevention and Response Victim Advocate (SAPR VA or VA)	NR (8.95–55.38)	NR (11.41–41.81)	NR (6.24–79.70)	NR (NR)	NR (NR)
A Safe Helpline or other hotline counselor	NR (1.62–53.35)	3.30% (0.09–16.61)	NR (2.92–79.25)	NR (NR)	NR (NR)
A medical professional (for example, a doctor or nurse)	NR (4.12–53.32)	10.39% (3.07–23.92)	NR (4.91–79.42)	NR (NR)	NR (NR)
A chaplain or religious leader	NR (5.16–53.83)	NR (4.60–32.79)	NR (4.21–79.31)	NR (NR)	NR (NR)
A counselor, therapist, or psychologist	NR (8.69–55.36)	NR (7.64–36.55)	NR (6.55–79.78)	NR (NR)	NR (NR)
A Special Victims' Counsel or Victims' Legal Counsel	NR (2.70–53.15)	7.31% (1.58–19.68)	NR (4.55–79.38)	NR (NR)	NR (NR)
Some other military lawyer (for example, a JAG officer)	NR (1.47–53.51)	7.42% (1.60–19.97)	NR (2.00–79.42)	NR (NR)	NR (NR)
A supervisor or someone above you in your chain of command	NR (12.11–57.48)	NR (8.47–37.78)	NR (14.91–82.50)	NR (NR)	NR (NR)
An officer or NCO outside of your chain of command	NR (5.30–55.31)	3.62% (0.28–14.29)	NR (4.55–79.39)	NR (NR)	NR (NR)
Someone in military law enforcement	NR (2.28–53.24)	6.93% (1.29–19.80)	NR (2.92–79.26)	NR (NR)	NR (NR)
Civilian law enforcement authority	NR (1.13–53.78)	0.00% (0.00–9.32)	NR (2.92–79.26)	NR (NR)	NR (NR)
Civilian rape crisis center or other sexual assault advocacy group	NR (0.87–54.07)	0.79% (0.00–9.70)	NR (2.00–79.42)	NR (NR)	NR (NR)

NOTE: Includes estimates for active-component Coast Guard members. 95-percent confidence intervals for each estimate are included in parentheses.

Too few warrant officers were included in the sample to break them out as a separate pay grade. For the purposes of this table, warrant officers have been included with the E5–E9 category.

NR = Not reportable.

A.24. Percentage of members who had a forensic exam following the single or most serious assault among members who experienced a penetrative sexual assault in the past year and told someone about it

Table A.24.a
Percentage of members who had a forensic exam following the single or most serious assault among members who experienced a penetrative sexual assault in the past year and told someone about it, by gender

SAFU21: Did you receive a sexual assault forensic exam or "rape exam"? This is often given by medical personnel to collect evidence about a sexual assault and could be either civilian or military.

Total	Men	Women
NR	NR	NR
(7.79–90.00)	(NR)	(1.81–33.57)

NOTE: Includes estimates for active-component Coast Guard members. 95-percent confidence intervals for each estimate are included in parentheses.

NR = Not reportable.

Table A.24.b
Percentage of members who had a forensic exam following the single or most serious assault among members who experienced a penetrative sexual assault in the past year and told someone about it, by service

SAFU21: Did you receive a sexual assault forensic exam or "rape exam"? This is often given by medical personnel to collect evidence about a sexual assault and could be either civilian or military.

Total DoD	Army	Navy	Air Force	Marine Corps	Coast Guard
28.88%	26.43%	18.47%	14.62%	NR	NR
(17.69–42.34)	(17.25–37.39)	(10.83–28.44)	(9.39–21.30)	(21.40–85.98)	(7.79–90.00)

NOTE: Includes estimates for active-component DoD and Coast Guard members. 95-percent confidence intervals for each estimate are included in parentheses.

NR = Not reportable.

Table A.24.c
Percentage of members who had a forensic exam following the single or most serious assault among members who experienced a penetrative sexual assault in the past year and told someone about it, by pay grade

SAFU21: Did you receive a sexual assault forensic exam or "rape exam"? This is often given by medical personnel to collect evidence about a sexual assault and could be either civilian or military.

Total	E1–E4	E5–E9	O1–O3	O4–O6
NR	NR	NR	NR	NR
(7.79–90.00)	(NR)	(NR)	(NR)	(NR)

NOTE: Includes estimates for active-component Coast Guard members. 95-percent confidence intervals for each estimate are included in parentheses.

Too few warrant officers were included in the sample to break them out as a separate pay grade. For the purposes of this table, warrant officers have been included with the E5–E9 category.

NR = Not reportable.

A.25. Satisfaction with treatment following the single or most serious assault among members who experienced a sexual assault in the past year

Table A.25
Satisfaction with treatment following the single or most serious assault among members who experienced a sexual assault in the past year, by gender

SAFU22: How satisfied have you been with your treatment by:

	Total	Men	Women
The Sexual Assault Response Coordinator (SARC)?			
Very dissatisfied	NR (9.23–95.94)	NR (NR)	NR (NR)
Dissatisfied	NR (NR)	NR (NR)	NR (NR)
Neither satisfied nor dissatisfied	NR (NR)	NR (NR)	NR (NR)
Satisfied	NR (2.32–58.14)	NR (NR)	NR (NR)
Very satisfied	NR (2.20–63.33)	NR (NR)	NR (NR)
Sexual Assault Prevention and Response Victim Advocates (SAPR VA or VA)?			
Very dissatisfied	NR (9.91–94.43)	NR (NR)	NR (0.28–42.48)
Dissatisfied	NR (NR)	NR (NR)	NR (NR)
Neither satisfied nor dissatisfied	NR (NR)	NR (NR)	NR (NR)
Satisfied	NR (1.91–46.34)	NR (NR)	NR (11.77–65.95)
Very satisfied	NR (4.08–70.52)	NR (NR)	NR (25.35–82.06)
Counselors, therapists or psychologists?			
Very dissatisfied	NR (6.48–94.64)	NR (NR)	NR (0.00–24.63)
Dissatisfied	NR (NR)	NR (NR)	NR (NR)
Neither satisfied nor dissatisfied	NR (1.69–49.62)	NR (NR)	NR (9.92–65.68)
Satisfied	NR (2.06–53.74)	NR (NR)	NR (12.67–69.57)
Very satisfied	NR (1.66–41.63)	NR (NR)	NR (8.62–56.95)

Table A.25—Continued

	Total	Men	Women
Supervisors or people in your chain of command?			
Very dissatisfied	NR (14.76–92.53)	NR (NR)	NR (2.09–72.53)
Dissatisfied	NR (NR)	NR (NR)	NR (NR)
Neither satisfied nor dissatisfied	NR (4.10–59.69)	NR (NR)	NR (13.69–74.87)
Satisfied	NR (1.00–33.14)	NR (NR)	NR (4.50–49.48)
Very satisfied	NR (0.72–31.76)	NR (NR)	NR (1.02–37.81)

NOTE: Includes estimates for active-component Coast Guard members. 95-percent confidence intervals for each estimate are included in parentheses.

Some questions that were asked received fewer than 15 respondents, resulting in results that are not reportable. Those questions include:
- Safe Helpline or other hotline counselors?
- Medical professional(s)? For example, a doctor or nurse.
- Chaplains or religious leaders?
- Special Victims' Counsel or Victims' Legal Counsel?
- Other military lawyers (for example, a JAG officer)?
- Officers or NCOs outside of your chain of command?
- Military law enforcement personnel?
- Civilian law enforcement personnel?
- Civilian rape crisis center or other sexual assault advocacy group?
- The sexual assault forensic exam?

NR = Not reportable.

A.26. Percentage of members who officially reported the single or most serious assault among members who experienced a sexual assault in the past year

Table A.26.a
Percentage of members who officially reported the single or most serious assault among members who experienced a sexual assault in the past year, by gender

SAFU23: Did you officially report this unwanted event to the military? This could have been either a restricted or unrestricted report.

Note: The following definition was kept on screen for items SAFU23, SAFU24, and SAFU25:

- DoD provides two types of sexual assault reports.
- Restricted reports allow people to get information, collect evidence, and receive medical treatment and counseling without starting an official investigation of the assault.
- Unrestricted reports start *an official investigation* in addition to allowing the services available in restricted reporting.

Total	Men	Women
NR	NR	18.27%
(8.28–54.65)	(NR)	(9.44–30.43)

NOTE: Includes estimates for active-component Coast Guard members. 95-percent confidence intervals for each estimate are included in parentheses.

NR = Not reportable.

Table A.26.b
Percentage of members who officially reported the single or most serious assault among members who experienced a sexual assault in the past year, by service

SAFU23: Did you officially report this unwanted event to the military? This could have been either a restricted or unrestricted report.

Total DoD	Army	Navy	Air Force	Marine Corps	Coast Guard
15.26%	19.04%	9.25%	16.63%	NR	NR
(12.24–18.70)	(14.87–23.79)	(6.44–12.77)	(13.26–20.46)	(6.61–39.88)	(8.28–54.65)

NOTE: Includes estimates for active-component DoD and Coast Guard members. 95-percent confidence intervals for each estimate are included in parentheses.

NR = Not reportable.

Table A.26.c
Percentage of members who officially reported the single or most serious assault among members who experienced a sexual assault in the past year, by pay grade

SAFU23: Did you officially report this unwanted event to the military? This could have been either a restricted or unrestricted report.

Total	E1–E4	E5–E9	O1–O3	O4–O6
NR	NR	NR	NR	NR
(8.28–54.65)	(10.60–40.67)	(5.98–78.87)	(NR)	(NR)

NOTE: Includes estimates for active-component Coast Guard members. 95-percent confidence intervals for each estimate are included in parentheses.

Too few warrant officers were included in the sample to break them out as a separate pay grade. For the purposes of this table, warrant officers have been included with the E5–E9 category.

NR = Not reportable.

A.27. Type of official report among members who experienced a sexual assault in the past year and made an official report to the military

Table A.27.a
Type of official report among members who experienced a sexual assault in the past year and made an official report to the military, by gender

SAFU24: Did you make...Select one

	Total	Men	Women
Only a restricted report?	NR (0.22–33.33)	NR (NR)	NR (NR)
Only an unrestricted report?	NR (4.11–73.63)	NR (NR)	NR (NR)
A restricted report that turned into an unrestricted report?	NR (10.99–94.93)	NR (NR)	NR (NR)
Or were you not sure what type of report it was?	NR (0.05–28.31)	NR (NR)	NR (NR)

NOTE: Includes estimates for active-component Coast Guard members. 95-percent confidence intervals for each estimate are included in parentheses.

NR = Not reportable.

Table A.27.b
Type of official report among members who experienced a sexual assault in the past year and made an official report to the military, by service

SAFU24: Did you make...Select one

	Total DoD	Army	Navy	Air Force	Marine Corps	Coast Guard
Only a restricted report?	26.92% (19.51–35.42)	31.24% (20.77–43.33)	29.64% (17.47–44.38)	26.27% (18.21–35.69)	NR (3.20–34.67)	NR (0.22–33.33)
Only an unrestricted report?	40.67% (31.24–50.63)	46.11% (35.33–57.17)	NR (23.32–52.26)	50.84% (39.73–61.89)	NR (5.56–56.01)	NR (4.11–73.63)
A restricted report that turned into an unrestricted report?	NR (10.77–40.29)	15.59% (9.92–22.82)	17.27% (9.01–28.68)	12.19% (6.83–19.60)	NR (15.46–91.23)	NR (10.99–94.93)
Or were you not sure what type of report it was?	9.22% (5.44–14.38)	7.06% (3.20–13.18)	NR (6.23–31.75)	10.70% (3.57–23.22)	NR (0.38–20.75)	NR (0.05–28.31)

NOTE: Includes estimates for active-component DoD and Coast Guard members. 95-percent confidence intervals for each estimate are included in parentheses.

NR = Not reportable.

Table A.27.c
Type of official report among members who experienced a sexual assault in the past year and made an official report to the military, by pay grade

SAFU24: Did you make...Select one

	Total	E1–E4	E5–E9	O1–O3	O4–O6
Only a restricted report?	NR (0.22–33.33)	NR (NR)	NR (NR)	NR (NR)	NR (NR)
Only an unrestricted report?	NR (4.11–73.63)	NR (NR)	NR (NR)	NR (NR)	NR (NR)
A restricted report that turned into an unrestricted report?	NR (10.99–94.93)	NR (NR)	NR (NR)	NR (NR)	NR (NR)
Or were you not sure what type of report it was?	NR (0.05–28.31)	NR (NR)	NR (NR)	NR (NR)	NR (NR)

NOTE: Includes estimates for active-component Coast Guard members. 95-percent confidence intervals for each estimate are included in parentheses.

Too few warrant officers were included in the sample to break them out as a separate pay grade. For the purposes of this table, warrant officers have been included with the E5–E9 category.

NR = Not reportable.

A.28. Reporting preference among members who experienced a sexual assault in the past year and made an unrestricted report or a restricted report that turned into an unrestricted report

Table A.28.a
Reporting preference among members who experienced a sexual assault in the past year and made an unrestricted report or a restricted report that turned into an unrestricted report, by gender

SAFU25: Was an unrestricted report what you preferred?

	Total	Men	Women
Yes, that's what you wanted	NR (NR)	NR (NR)	NR (NR)
No, you wanted a restricted report, but could not	NR (NR)	NR (NR)	NR (NR)

NOTE: Includes estimates for active-component Coast Guard members. 95-percent confidence intervals for each estimate are included in parentheses.

NR = Not reportable.

Table A.28.b
Reporting preference among members who experienced a sexual assault in the past year and made an unrestricted report or a restricted report that turned into an unrestricted report, by service

SAFU25: Was an unrestricted report what you preferred?

	Total DoD	Army	Navy	Air Force	Marine Corps	Coast Guard
Yes, that's what you wanted	77.29% (67.50–85.30)	75.05% (64.26–83.96)	NR (48.22–84.57)	78.86% (67.55–87.66)	NR (62.61–98.19)	NR (NR)
No, you wanted a restricted report, but could not	22.71% (14.70–32.50)	24.95% (16.04–35.74)	NR (15.43–51.78)	21.14% (12.34–32.45)	NR (1.81–37.39)	NR (NR)

NOTE: Includes estimates for active-component DoD and Coast Guard members. 95-percent confidence intervals for each estimate are included in parentheses.

NR = Not reportable.

Table A.28.c
Reporting preference among members who experienced a sexual assault in the past year and made an unrestricted report or a restricted report that turned into an unrestricted report, by pay grade

SAFU25: Was an unrestricted report what you preferred?

	Total	E1–E4	E5–E9	O1–O3	O4–O6
Yes, that's what you wanted	NR (NR)	NR (NR)	NR (NR)	NR (NR)	NR (NR)
No, you wanted a restricted report, but could not	NR (NR)	NR (NR)	NR (NR)	NR (NR)	NR (NR)

NOTE: Includes estimates for active-component Coast Guard members. 95-percent confidence intervals for each estimate are included in parentheses.

Too few warrant officers were included in the sample to break them out as a separate pay grade. For the purposes of this table, warrant officers have been included with the E5–E9 category.

NR = Not reportable.

A.29. Percentage of members who were interviewed by an investigator about the single or most serious assault among members who experienced a sexual assault in the past year

Table A.29.a
Percentage of members who were interviewed by an investigator about the single or most serious assault among members who experienced a sexual assault in the past year, by gender

SAFU26: Have military police or criminal investigators interviewed you about the case?

	Total	Men	Women
Yes	NR (5.52–53.31)	NR (NR)	12.73% (5.74–23.38)
No	NR (46.69–94.48)	NR (NR)	87.27% (76.62–94.26)

NOTE: Includes estimates for active-component Coast Guard members. 95-percent confidence intervals for each estimate are included in parentheses.

NR = Not reportable.

Table A.29.b
Percentage of members who were interviewed by an investigator about the single or most serious assault among members who experienced a sexual assault in the past year, by service

SAFU26: Have military police or criminal investigators interviewed you about the case?

	Total DoD	Army	Navy	Air Force	Marine Corps	Coast Guard
Yes	7.79% (6.40–9.36)	9.92% (7.38–12.97)	5.82% (3.71–8.63)	10.27% (7.68–13.35)	5.27% (2.62–9.33)	NR (5.52–53.31)
No	92.21% (90.64–93.60)	90.08% (87.03–92.62)	94.18% (91.37–96.29)	89.73% (86.65–92.32)	94.73% (90.67–97.38)	NR (46.69–94.48)

NOTE: Includes estimates for active-component DoD and Coast Guard members. 95-percent confidence intervals for each estimate are included in parentheses.

NR = Not reportable.

Table A.29.c
Percentage of members who were interviewed by an investigator about the single or most serious assault among members who experienced a sexual assault in the past year, by pay grade

SAFU26: Have military police or criminal investigators interviewed you about the case?

	Total	E1–E4	E5–E9	O1–O3	O4–O6
Yes	NR (5.52–53.31)	14.18% (5.13–29.04)	NR (5.62–78.80)	NR (NR)	NR (NR)
No	NR (46.69–94.48)	85.82% (70.96–94.87)	NR (21.20–94.38)	NR (NR)	NR (NR)

NOTE: Includes estimates for active-component Coast Guard members. 95-percent confidence intervals for each estimate are included in parentheses.

Too few warrant officers were included in the sample to break them out as a separate pay grade. For the purposes of this table, warrant officers have been included with the E5–E9 category.

NR = Not reportable.

A.30. Whether or not a suspect was arrested for the single or most serious assault among members who experienced a sexual assault in the past year

Table A.30.a
Whether or not a suspect was arrested for the single or most serious assault among members who experienced a sexual assault in the past year, by gender

SAFU27a: Has a suspect been arrested or charged with a crime?

	Total	Men	Women
Yes	1.07%	NR	1.69%
	(0.02–6.10)	(NR)	(0.04–9.02)
No	93.92%	NR	92.46%
	(86.03–98.09)	(NR)	(82.49–97.73)
Do not know	5.02%	NR	5.85%
	(1.36–12.46)	(NR)	(1.40–15.26)

NOTE: Includes estimates for active-component Coast Guard members. 95-percent confidence intervals for each estimate are included in parentheses.

NR = Not reportable.

Table A.30.b
Whether or not a suspect was arrested for the single or most serious assault among members who experienced a sexual assault in the past year, by service

SAFU27a: Has a suspect been arrested or charged with a crime?

	Total DoD	Army	Navy	Air Force	Marine Corps	Coast Guard
Yes	2.08%	2.84%	1.14%	2.83%	1.88%	1.07%
	(1.51–2.80)	(1.76–4.32)	(0.42–2.48)	(1.41–5.02)	(0.49–4.86)	(0.02–6.10)
No	95.60%	94.13%	97.24%	93.84%	96.66%	93.92%
	(94.39–96.61)	(91.72–96.01)	(95.01–98.65)	(91.10–95.94)	(93.15–98.68)	(86.03–98.09)
Do not know	2.32%	3.03%	1.62%	3.33%	1.46%	5.02%
	(1.54–3.35)	(1.58–5.21)	(0.51–3.80)	(1.90–5.37)	(0.30–4.25)	(1.36–12.46)

NOTE: Includes estimates for active-component DoD and Coast Guard members. 95-percent confidence intervals for each estimate are included in parentheses.

Table A.30.c
Whether or not a suspect was arrested for the single or most serious assault among members who experienced a sexual assault in the past year, by pay grade

SAFU27a: Has a suspect been arrested or charged with a crime?

	Total	E1–E4	E5–E9	O1–O3	O4–O6
Yes	1.07%	2.97%	NR	NR	NR
	(0.02–6.10)	(0.08–15.10)	(0.00–38.25)	(NR)	(NR)
No	93.92%	NR	97.30%	NR	NR
	(86.03–98.09)	(70.84–95.81)	(82.59–99.98)	(NR)	(NR)
Do not know	5.02%	NR	2.70%	NR	NR
	(1.36–12.46)	(2.60–25.46)	(0.02–17.41)	(NR)	(NR)

NOTE: Includes estimates for active-component Coast Guard members. 95-percent confidence intervals for each estimate are included in parentheses.

Too few warrant officers were included in the sample to break them out as a separate pay grade. For the purposes of this table, warrant officers have been included with the E5–E9 category.

NR = Not reportable.

A.31. Reasons for reporting the single or most serious assault among members who experienced a sexual assault in the past year and made an official report to the military

Table A.31.a
Reasons for reporting the single or most serious assault among members who experienced a sexual assault in the past year and made an official report to the military, by gender

SAFU29: What were your reasons for reporting the event to a military authority?

	Total	Men	Women
Someone else made you report it or reported it themselves.	NR (40.18–97.20)	NR (NR)	NR (NR)
To stop the offender(s) from hurting you again.	NR (21.67–95.61)	NR (NR)	NR (NR)
To stop the offender(s) from hurting others.	NR (34.13–96.81)	NR (NR)	NR (NR)
It was your civic/military duty to report it.	NR (25.50–95.92)	NR (NR)	NR (NR)
To punish the offender(s).	NR (35.13–96.63)	NR (NR)	NR (NR)
To discourage other potential offenders.	NR (19.22–95.36)	NR (NR)	NR (NR)
To get medical assistance.	NR (8.79–94.87)	NR (NR)	NR (NR)
To get mental health assistance.	NR (27.16–96.07)	NR (NR)	NR (NR)
To stop rumors.	NR (12.11–94.96)	NR (NR)	NR (NR)
Someone you told encouraged you to report.	NR (28.81–96.10)	NR (NR)	NR (NR)

NOTE: Includes estimates for active-component Coast Guard members. 95-percent confidence intervals for each estimate are included in parentheses.

NR = Not reportable.

Table A.31.b
Reasons for reporting the single or most serious assault among members who experienced
a sexual assault in the past year and made an official report to the military, by service

SAFU29: What were your reasons for reporting the event to a military authority?

	Total DoD	Army	Navy	Air Force	Marine Corps	Coast Guard
Someone else made you report it or reported it themselves.	26.53% (19.30–34.81)	28.13% (18.85–39.01)	NR (20.00–49.44)	26.64% (18.44–36.23)	NR (2.72–36.75)	NR (40.18–97.20)
To stop the offender(s) from hurting you again.	35.80% (27.16–45.18)	41.49% (31.05–52.52)	36.16% (23.12–50.88)	38.28% (27.52–49.96)	NR (4.12–47.16)	NR (21.67–95.61)
To stop the offender(s) from hurting others.	49.54% (38.45–60.66)	53.61% (42.46–64.50)	50.84% (35.97–65.61)	61.28% (50.64–71.19)	NR (6.00–64.90)	NR (34.13–96.81)
It was your civic/ military duty to report it.	29.52% (21.70–38.32)	38.35% (27.71–49.87)	22.58% (12.75–35.27)	29.38% (19.32–41.17)	NR (2.87–42.45)	NR (25.50–95.92)
To punish the offender(s).	NR (17.26–45.30)	24.65% (16.55–34.31)	19.23% (8.87–34.03)	20.32% (11.38–32.07)	NR (23.14–92.71)	NR (35.13–96.63)
To discourage other potential offenders.	20.66% (14.65–27.81)	25.08% (16.71–35.08)	16.17% (6.90–30.20)	25.69% (15.18–38.74)	NR (2.03–31.48)	NR (19.22–95.36)
To get medical assistance.	23.30% (17.10–30.48)	24.04% (16.60–32.84)	29.65% (17.76–43.98)	22.79% (13.58–34.42)	NR (2.80–37.13)	NR (8.79–94.87)
To get mental health assistance.	30.31% (22.93–38.51)	31.48% (23.11–40.85)	31.90% (19.65–46.31)	31.40% (21.28–43.01)	NR (5.32–56.78)	NR (27.16–96.07)
To stop rumors.	NR (8.85–39.55)	12.85% (7.56–19.98)	7.45% (2.56–16.20)	17.53% (7.74–31.97)	NR (21.83–92.59)	NR (12.11–94.96)
Someone you told encouraged you to report.	43.29% (33.40–53.59)	41.10% (31.23–51.53)	NR (30.46–60.60)	58.08% (47.01–68.57)	NR (7.15–74.48)	NR (28.81–96.10)

NOTE: Includes estimates for active-component DoD and Coast Guard members. 95-percent confidence
intervals for each estimate are included in parentheses.

NR = Not reportable.

Table A.31.c
Reasons for reporting the single or most serious assault among members who experienced a sexual assault in the past year and made an official report to the military, by pay grade

SAFU29: What were your reasons for reporting the event to a military authority?

	Total	E1–E4	E5–E9	O1–O3	O4–O6
Someone else made you report it or reported it themselves.	NR (40.18–97.20)	NR (NR)	NR (NR)	NR (NR)	NR (NR)
To stop the offender(s) from hurting you again.	NR (21.67–95.61)	NR (NR)	NR (NR)	NR (NR)	NR (NR)
To stop the offender(s) from hurting others.	NR (34.13–96.81)	NR (NR)	NR (NR)	NR (NR)	NR (NR)
It was your civic/military duty to report it.	NR (25.50–95.92)	NR (NR)	NR (NR)	NR (NR)	NR (NR)
To punish the offender(s).	NR (35.13–96.63)	NR (NR)	NR (NR)	NR (NR)	NR (NR)
To discourage other potential offenders.	NR (19.22–95.36)	NR (NR)	NR (NR)	NR (NR)	NR (NR)
To get medical assistance.	NR (8.79–94.87)	NR (NR)	NR (NR)	NR (NR)	NR (NR)
To get mental health assistance.	NR (27.16–96.07)	NR (NR)	NR (NR)	NR (NR)	NR (NR)
To stop rumors.	NR (12.11–94.96)	NR (NR)	NR (NR)	NR (NR)	NR (NR)
Someone you told encouraged you to report.	NR (28.81–96.10)	NR (NR)	NR (NR)	NR (NR)	NR (NR)

NOTE: Includes estimates for active-component Coast Guard members. 95-percent confidence intervals for each estimate are included in parentheses.

Too few warrant officers were included in the sample to break them out as a separate pay grade. For the purposes of this table, warrant officers have been included with the E5–E9 category.

NR = Not reportable.

A.32. Reasons for not reporting the single or most serious assault among members who experienced a sexual assault in the past year and did not officially report it to the military

Table A.32.a
Reasons for not reporting the single or most serious assault among members who experienced a sexual assault in the past year and did not officially report it to the military, by gender

SAFU30: What were your reasons for not reporting the event to a military authority?

	Total	Men	Women
Someone else already reported it.	0.89% (0.00–6.72)	NR (NR)	1.26% (0.01–7.90)
You thought it was not serious enough to report.	NR (31.38–67.18)	NR (NR)	NR (38.18–69.04)
You did not want more people to know.	NR (30.99–66.46)	NR (NR)	NR (53.25–81.61)
You did not want people to see you as weak.	26.06% (14.46–40.75)	NR (NR)	NR (21.16–50.63)
You did not want people to think you were gay/lesbian/bisexual/transgender.	1.68% (0.04–9.08)	NR (NR)	2.37% (0.06–12.47)
You wanted to forget about it and move on.	NR (38.38–76.84)	NR (NR)	80.82% (67.77–90.28)
You did not know how to report it.	3.30% (0.51–10.52)	NR (NR)	2.66% (0.19–11.00)
Someone told you not to report it.	0.88% (0.00–6.70)	NR (NR)	1.24% (0.01–7.86)
You did not think your report would be kept confidential.	28.38% (16.31–43.25)	NR (NR)	NR (25.80–55.79)
You did not think anything would be done.	NR (32.53–68.33)	NR (NR)	NR (27.73–59.61)
You did not think you would be believed.	21.46% (11.15–35.30)	NR (NR)	NR (15.17–42.57)
You did not trust the process would be fair.	NR (20.72–57.96)	NR (NR)	NR (20.12–52.73)
You felt partially to blame.	NR (18.63–47.05)	NR (NR)	NR (29.70–60.36)
You thought other people would blame you.	NR (25.76–58.57)	NR (NR)	NR (42.46–73.55)
You thought you might get in trouble for something you did (for example, underage drinking or fraternization).	12.46% (5.48–23.21)	NR (NR)	17.61% (8.34–30.90)
You thought you might be labeled as a troublemaker.	NR (18.22–49.00)	NR (NR)	NR (28.06–59.98)

Table A.32.a—Continued

	Total	Men	Women
You thought it might hurt your performance evaluation/fitness report.	16.32% (7.69–28.84)	NR (NR)	21.07% (10.36–35.80)
You thought it might hurt your career.	NR (17.88–53.93)	NR (NR)	NR (19.71–48.18)
You did not want to hurt the person's career or family.	NR (21.81–53.69)	NR (NR)	NR (36.24–67.22)
You were worried about retaliation by the person(s) who did it.	NR (13.64–42.17)	NR (NR)	NR (20.16–52.07)
You were worried about retaliation by supervisor or someone in your chain of command.	NR (8.21–47.03)	NR (NR)	NR (6.11–37.02)
You were worried about retaliation by your military co-workers or peers.	NR (24.35–61.00)	NR (NR)	NR (28.10–59.97)
You took other actions to handle the situation.	NR (29.35–65.10)	NR (NR)	NR (28.77–59.38)

NOTE: Includes estimates for active-component Coast Guard members. 95-percent confidence intervals for each estimate are included in parentheses.

Some response options that appear in Table A.32.b do not appear here because they were not selected by any Coast Guard respondent.

NR = Not reportable.

Table A.32.b
Reasons for not reporting the single or most serious assault among members who experienced a sexual assault in the past year and did not officially report it to the military, by service

SAFU30: What were your reasons for not reporting the event to a military authority?

	Total DoD	Army	Navy	Air Force	Marine Corps	Coast Guard
You reported it to civilian authorities/ law enforcement.	2.26% (0.42–6.77)	3.80% (0.16–17.55)	0.26% (0.01–1.31)	0.91% (0.26–2.26)	NR (0.30–21.80)	0.00% (0.00–9.21)
Someone else already reported it.	2.71% (0.85–6.36)	5.37% (0.82–16.80)	0.91% (0.21–2.49)	1.36% (0.52–2.88)	2.49% (0.52–7.11)	0.89% (0.00–6.72)
You thought it was not serious enough to report.	47.87% (41.90–53.88)	47.33% (37.84–56.97)	47.96% (36.87–59.20)	55.16% (48.65–61.55)	NR (26.48–59.96)	NR (31.38–67.18)
You did not want more people to know.	56.36% (50.35–62.24)	50.53% (41.11–59.92)	61.76% (50.26–72.37)	52.16% (45.51–58.76)	NR (41.95–74.79)	NR (30.99–66.46)
You did not want people to see you as weak.	39.68% (33.41–46.22)	39.51% (29.77–49.90)	39.35% (28.01–51.60)	27.30% (21.54–33.68)	NR (34.03–69.05)	26.06% (14.46–40.75)
You did not want people to think you were gay/lesbian/ bisexual/transgender.	15.71% (9.68–23.52)	15.52% (7.25–27.66)	NR (6.89–33.47)	4.59% (1.77–9.50)	NR (5.83–46.77)	1.68% (0.04–9.08)
You wanted to forget about it and move on.	68.40% (63.24–73.24)	71.86% (63.72–79.08)	69.30% (59.92–77.64)	64.85% (58.11–71.17)	NR (43.30–76.58)	NR (38.38–76.84)
You did not know how to report it.	6.55% (3.54–10.91)	10.86% (3.81–23.00)	5.80% (2.67–10.78)	2.17% (0.91–4.32)	2.19% (0.53–5.87)	3.30% (0.51–10.52)
Someone told you not to report it.	8.88% (3.97–16.61)	11.24% (4.08–23.33)	NR (0.99–29.44)	2.82% (1.20–5.57)	8.55% (1.93–22.39)	0.88% (0.00–6.70)
You did not think your report would be kept confidential.	30.52% (24.44–37.15)	36.83% (27.03–47.50)	30.15% (19.06–43.24)	24.35% (19.30–29.99)	NR (9.14–40.66)	28.38% (16.31–43.25)
You did not think anything would be done.	39.89% (33.80–46.23)	42.46% (32.93–52.41)	40.14% (28.82–52.29)	32.48% (26.40–39.03)	NR (23.09–57.99)	NR (32.53–68.33)
You did not think you would be believed.	25.51% (19.72–32.01)	33.47% (23.86–44.21)	24.51% (13.79–38.18)	22.70% (17.04–29.21)	11.88% (5.75–20.99)	21.46% (11.15–35.30)
You did not trust the process would be fair.	32.06% (26.11–38.47)	40.43% (30.74–50.70)	32.98% (21.94–45.61)	23.11% (18.25–28.56)	17.60% (8.30–30.95)	NR (20.72–57.96)
You felt partially to blame.	29.37% (24.84–34.23)	32.98% (24.37–42.52)	28.22% (20.88–36.51)	33.21% (27.77–39.01)	20.70% (12.70–30.81)	NR (18.63–47.05)
You thought other people would blame you.	34.78% (28.49–41.49)	31.68% (23.04–41.36)	40.80% (28.68–53.79)	30.96% (25.74–36.57)	NR (14.93–47.07)	NR (25.76–58.57)
You thought you might get in trouble for something you did (for example, underage drinking or fraternization).	17.68% (13.19–22.93)	18.98% (11.16–29.14)	15.95% (10.56–22.69)	9.01% (5.98–12.89)	NR (10.75–49.08)	12.46% (5.48–23.21)

Table A.32.b—Continued

	Total DoD	Army	Navy	Air Force	Marine Corps	Coast Guard
You thought you might be labeled as a troublemaker.	27.38% (21.88–33.43)	31.00% (22.98–39.95)	29.30% (18.41–42.25)	19.56% (15.03–24.75)	20.41% (10.53–33.81)	NR (18.22–49.00)
You thought it might hurt your performance evaluation/fitness report.	23.20% (17.49–29.74)	25.71% (17.30–35.67)	25.09% (14.30–38.73)	16.54% (11.77–22.29)	NR (6.06–37.09)	16.32% (7.69–28.84)
You thought it might hurt your career.	37.09% (31.15–43.34)	45.58% (36.11–55.28)	38.19% (26.97–50.40)	32.16% (26.36–38.39)	18.43% (9.15–31.42)	NR (17.88–53.93)
You did not want to hurt the person's career or family.	35.48% (29.67–41.62)	33.33% (24.60–43.00)	36.95% (26.35–48.57)	33.95% (28.21–40.07)	NR (20.73–57.56)	NR (21.81–53.69)
You were worried about retaliation by the person(s) who did it.	32.19% (26.21–38.64)	38.01% (28.26–48.53)	31.60% (20.52–44.46)	26.34% (21.01–32.25)	NR (13.52–40.16)	NR (13.64–42.17)
You were worried about retaliation by supervisor or someone in your chain of command.	22.51% (16.66–29.28)	29.46% (19.78–40.72)	21.19% (10.59–35.66)	12.93% (9.14–17.58)	17.99% (8.34–31.89)	NR (8.21–47.03)
You were worried about retaliation by your military co-workers or peers.	28.22% (22.30–34.74)	34.68% (25.01–45.37)	30.20% (19.05–43.37)	20.12% (15.75–25.09)	14.63% (7.58–24.56)	NR (24.35–61.00)
You took other actions to handle the situation.	38.08% (32.63–43.77)	41.40% (32.09–51.18)	35.94% (26.69–46.02)	40.99% (34.54–47.68)	NR (19.46–50.25)	NR (29.35–65.10)

NOTE: Includes estimates for active-component DoD and Coast Guard members. 95-percent confidence intervals for each estimate are included in parentheses.

NR = Not reportable.

Table A.32.c
Reasons for not reporting the single or most serious assault among members who experienced a sexual assault in the past year and did not officially report it to the military, by pay grade

SAFU30: What were your reasons for not reporting the event to a military authority?

	Total	E1–E4	E5–E9	O1–O3	O4–O6
Someone else already reported it.	0.89% (0.00–6.72)	2.36% (0.03–14.41)	NR (0.00–31.46)	NR (NR)	NR (NR)
You thought it was not serious enough to report.	NR (31.38–67.18)	NR (30.84–70.79)	NR (10.66–62.57)	NR (NR)	NR (NR)
You did not want more people to know.	NR (30.99–66.46)	NR (44.37–83.52)	NR (11.52–64.02)	NR (NR)	NR (NR)
You did not want people to see you as weak.	26.06% (14.46–40.75)	NR (18.09–57.56)	NR (6.51–43.46)	NR (NR)	NR (NR)
You did not want people to think you were gay/lesbian/bisexual/transgender.	1.68% (0.04–9.08)	NR (0.14–21.60)	NR (0.00–31.46)	NR (NR)	NR (NR)
You wanted to forget about it and move on.	NR (38.38–76.84)	NR (71.45–96.89)	NR (14.60–69.99)	NR (NR)	NR (NR)
You did not know how to report it.	3.30% (0.51–10.52)	5.00% (0.39–19.35)	NR (0.03–21.02)	NR (NR)	NR (NR)
Someone told you not to report it.	0.88% (0.00–6.70)	0.00% (0.00–12.06)	NR (0.00–19.01)	NR (NR)	NR (NR)
You did not think your report would be kept confidential.	28.38% (16.31–43.25)	NR (27.32–67.11)	NR (2.73–34.44)	NR (NR)	NR (NR)
You did not think anything would be done.	NR (32.53–68.33)	NR (28.53–68.57)	NR (19.61–80.38)	NR (NR)	NR (NR)
You did not think you would be believed.	21.46% (11.15–35.30)	NR (14.04–52.08)	NR (2.39–33.47)	NR (NR)	NR (NR)
You did not trust the process would be fair.	NR (20.72–57.96)	NR (20.23–60.70)	NR (16.78–78.73)	NR (NR)	NR (NR)
You felt partially to blame.	NR (18.63–47.05)	NR (31.10–71.06)	NR (4.63–40.58)	NR (NR)	NR (NR)
You thought other people would blame you.	NR (25.76–58.57)	NR (51.03–86.10)	NR (6.47–44.72)	NR (NR)	NR (NR)
You thought you might get in trouble for something you did (for example, underage drinking or fraternization).	12.46% (5.48–23.21)	NR (9.72–43.08)	NR (0.53–26.29)	NR (NR)	NR (NR)
You thought you might be labeled as a troublemaker.	NR (18.22–49.00)	NR (23.70–63.63)	NR (9.06–60.12)	NR (NR)	NR (NR)

Table A.32.c—Continued

	Total	E1–E4	E5–E9	O1–O3	O4–O6
You thought it might hurt your performance evaluation/fitness report.	16.32% (7.69–28.84)	NR (7.30–41.30)	NR (2.19–32.88)	NR (NR)	NR (NR)
You thought it might hurt your career.	NR (17.88–53.93)	NR (15.53–52.55)	NR (14.32–76.40)	NR (NR)	NR (NR)
You did not want to hurt the person's career or family.	NR (21.81–53.69)	NR (43.17–81.38)	NR (7.03–55.68)	NR (NR)	NR (NR)
You were worried about retaliation by the person(s) who did it.	NR (13.64–42.17)	NR (21.32–60.97)	NR (6.58–54.65)	NR (NR)	NR (NR)
You were worried about retaliation by supervisor or someone in your chain of command.	NR (8.21–47.03)	NR (0.17–25.38)	NR (14.24–77.21)	NR (NR)	NR (NR)
You were worried about retaliation by your military co-workers or peers.	NR (24.35–61.00)	NR (22.12–62.17)	NR (23.25–82.68)	NR (NR)	NR (NR)
You took other actions to handle the situation.	NR (29.35–65.10)	NR (25.17–65.06)	NR (10.80–67.11)	NR (NR)	NR (NR)

NOTE: Includes estimates for active-component Coast Guard members. 95-percent confidence intervals for each estimate are included in parentheses.

Too few warrant officers were included in the sample to break them out as a separate pay grade. For the purposes of this table, warrant officers have been included with the E5–E9 category.

Some response options that appear in Table A.32.b do not appear here because they were not selected by any Coast Guard respondent.

NR = Not reportable.

A.33. Main reason for not reporting the single or most serious assault among members who experienced a sexual assault in the past year and did not officially report it to the military

Table A.33.a
Main reason for not reporting the single or most serious assault among members who experienced a sexual assault in the past year and did not officially report it to the military, by gender

Derived variable:

For those who chose only one reason:

SAFU30: What were your reasons for not reporting the event to a military authority?

For those who chose more than one reason:

SAFU31: Which was the main reason for not reporting the event?

	Total	Male	Female
Someone else already reported it.	0.39% (0.00–5.88)	NR (NR)	0.55% (0.00–6.73)
You thought it was not serious enough to report.	8.75% (3.17–18.41)	NR (NR)	12.39% (4.79–24.75)
You did not want more people to know.	3.11% (0.43–10.32)	NR (NR)	4.40% (0.78–13.14)
You wanted to forget about it and move on.	NR (6.98–32.78)	NR (NR)	NR (11.02–42.63)
You did not think your report would be kept confidential.	3.20% (0.36–11.41)	NR (NR)	4.54% (0.54–15.59)
You did not think anything would be done.	1.46% (0.04–7.94)	NR (NR)	2.06% (0.05–10.95)
You did not think you would be believed.	2.63% (0.28–9.59)	NR (NR)	3.72% (0.42–13.09)
You did not trust the process would be fair.	3.47% (0.37–12.48)	NR (NR)	1.88% (0.05–10.02)
You felt partially to blame.	5.99% (1.50–15.33)	NR (NR)	8.48% (2.25–20.75)
You thought other people would blame you.	1.49% (0.04–8.11)	NR (NR)	2.11% (0.05–11.18)
You thought you might get in trouble for something you did (for example, underage drinking or fraternization).	1.82% (0.09–8.35)	NR (NR)	2.58% (0.22–10.11)
You thought you might be labeled as a troublemaker.	0.88% (0.00–6.77)	NR (NR)	1.24% (0.01–7.96)
You thought it might hurt your career.	NR (0.49–41.09)	NR (NR)	0.99% (0.00–7.52)

Table A.33.a—Continued

	Total	Male	Female
You did not want to hurt the person's career or family.	4.17% (0.75–12.44)	NR (NR)	5.90% (1.12–16.95)
You were worried about retaliation by the person(s) who did it.	2.01% (0.13–8.65)	NR (NR)	2.85% (0.23–11.21)
You were worried about retaliation by supervisor or someone in your chain of command.	1.42% (0.03–7.74)	NR (NR)	0.00% (0.00–7.05)
You took other actions to handle the situation.	NR (15.87–51.28)	NR (NR)	21.97% (11.03–36.78)

NOTE: Includes estimates for active-component Coast Guard members. 95-percent confidence intervals for each estimate are included in parentheses.

Some response options that appear in Table A.33.b do not appear here because they were not selected by any Coast Guard respondent.

NR = Not reportable.

Table A.33.b
Main reason for not reporting the single or most serious assault among members who experienced a sexual assault in the past year and did not officially report it to the military, by service

Derived variable:

For those who chose only one reason:

SAFU30: What were your reasons for not reporting the event to a military authority?

For those who chose more than one reason:

SAFU31: Which was the main reason for not reporting the event?

	Total DoD	Army	Navy	Air Force	Marine Corps	Coast Guard
You reported it to civilian authorities/ law enforcement.	0.04% (0.00–0.31)	0.00% (0.00–2.63)	0.00% (0.00–3.81)	0.31% (0.02–1.36)	0.00% (0.00–9.50)	0.00% (0.00–9.27)
Someone else already reported it.	0.38% (0.10–0.99)	0.32% (0.03–1.24)	0.44% (0.01–2.41)	0.76% (0.19–2.04)	0.00% (0.00–9.50)	0.39% (0.00–5.88)
You thought it was not serious enough to report.	18.12% (14.37–22.38)	13.49% (7.53–21.69)	17.68% (11.96–24.72)	23.50% (17.99–29.76)	NR (13.27–42.14)	8.75% (3.17–18.41)
You did not want more people to know.	5.90% (4.47–7.63)	6.28% (3.94–9.43)	5.10% (2.86–8.33)	8.86% (5.88–12.69)	4.57% (1.53–10.25)	3.11% (0.43–10.32)
You did not want people to see you as weak.	3.31% (1.29–6.84)	1.32% (0.31–3.59)	2.22% (0.70–5.17)	2.13% (0.09–10.04)	NR (1.87–35.74)	0.00% (0.00–9.27)
You did not want people to think you were gay/lesbian/ bisexual/transgender.	3.60% (1.22–8.03)	5.11% (0.64–17.14)	3.09% (0.47–9.91)	0.00% (0.00–1.29)	4.58% (0.29–18.86)	0.00% (0.00–9.27)
You wanted to forget about it and move on.	17.08% (12.10–23.09)	13.50% (9.00–19.17)	18.16% (8.94–31.13)	17.12% (13.41–21.38)	NR (7.09–47.20)	NR (6.98–32.78)
You did not know how to report it.	0.57% (0.07–2.02)	1.32% (0.08–5.90)	0.10% (0.00–1.04)	0.62% (0.02–3.38)	0.00% (0.00–9.50)	0.00% (0.00–9.27)
Someone told you not to report it.	0.21% (0.04–0.66)	0.37% (0.02–1.75)	0.00% (0.00–3.81)	0.15% (0.00–1.08)	0.46% (0.00–3.21)	0.00% (0.00–9.27)
You did not think your report would be kept confidential.	1.77% (0.88–3.16)	2.56% (1.11–4.98)	1.67% (0.29–5.16)	1.91% (0.26–6.42)	0.00% (0.00–9.50)	3.20% (0.36–11.41)
You did not think anything would be done.	6.15% (4.20–8.63)	7.46% (4.26–11.96)	5.38% (2.48–10.00)	5.41% (1.71–12.41)	5.78% (1.34–15.29)	1.46% (0.04–7.94)
You did not think you would be believed.	0.79% (0.42–1.35)	0.48% (0.09–1.49)	0.23% (0.01–1.26)	2.59% (1.25–4.73)	1.49% (0.23–4.87)	2.63% (0.28–9.59)
You did not trust the process would be fair.	1.92% (1.14–3.02)	2.42% (0.85–5.35)	1.64% (0.67–3.33)	3.23% (1.41–6.23)	0.32% (0.00–2.95)	3.47% (0.37–12.48)
You felt partially to blame.	5.74% (4.34–7.42)	4.36% (2.78–6.47)	6.38% (3.72–10.09)	6.89% (4.76–9.58)	6.27% (2.63–12.29)	5.99% (1.50–15.33)

Table A.33.b—Continued

	Total DoD	Army	Navy	Air Force	Marine Corps	Coast Guard
You thought other people would blame you.	0.90% (0.48–1.53)	0.80% (0.24–1.95)	1.03% (0.26–2.71)	1.11% (0.37–2.54)	0.58% (0.01–3.41)	1.49% (0.04–8.11)
You thought you might get in trouble for something you did (for example, underage drinking or fraternization).	1.98% (0.90–3.76)	1.58% (0.72–3.00)	2.84% (0.60–8.01)	1.23% (0.44–2.70)	1.22% (0.14–4.46)	1.82% (0.09–8.35)
You thought you might be labeled as a troublemaker.	1.05% (0.29–2.65)	1.78% (0.17–6.79)	0.94% (0.15–3.06)	0.30% (0.02–1.33)	0.27% (0.00–2.86)	0.88% (0.00–6.77)
You thought it might hurt your performance evaluation/fitness report.	0.36% (0.08–1.01)	0.40% (0.04–1.50)	0.58% (0.05–2.26)	0.00% (0.00–1.29)	0.00% (0.00–9.50)	0.00% (0.00–9.27)
You thought it might hurt your career.	2.45% (1.55–3.66)	3.09% (1.81–4.89)	2.10% (0.64–5.00)	2.99% (1.00–6.75)	1.41% (0.12–5.62)	NR (0.49–41.09)
You did not want to hurt the person's career or family.	5.72% (3.10–9.55)	10.28% (3.84–21.12)	3.91% (1.70–7.57)	3.23% (1.82–5.26)	1.97% (0.42–5.58)	4.17% (0.75–12.44)
You were worried about retaliation by the person(s) who did it.	1.50% (0.79–2.58)	1.71% (0.41–4.62)	1.12% (0.35–2.67)	1.87% (0.85–3.56)	1.73% (0.19–6.27)	2.01% (0.13–8.65)
You were worried about retaliation by supervisor or someone in your chain of command.	6.25% (1.83–14.84)	6.35% (1.09–18.81)	NR (1.38–30.26)	1.01% (0.31–2.40)	0.50% (0.01–3.28)	1.42% (0.03–7.74)
You were worried about retaliation by your military co-workers or peers.	1.36% (0.66–2.46)	2.56% (0.92–5.58)	0.73% (0.12–2.35)	0.74% (0.17–2.01)	0.74% (0.02–4.09)	0.00% (0.00–9.27)
You took other actions to handle the situation.	12.84% (9.91–16.28)	12.45% (8.42–17.51)	14.79% (8.97–22.44)	14.05% (9.97–19.01)	7.22% (2.78–14.82)	NR (15.87–51.28)

NOTE: Includes estimates for active-component DoD and Coast Guard members. 95-percent confidence intervals for each estimate are included in parentheses.

NR = Not reportable.

Table A.33.c

Main reason for not reporting the single or most serious assault among members who experienced a sexual assault in the past year and did not officially report it to the military, by pay grade

Derived variable:

For those who chose only one reason:

SAFU30: What were your reasons for not reporting the event to a military authority?

For those who chose more than one reason:

SAFU31: Which was the main reason for not reporting the event?

	Total	E1–E4	E5–E9	O1–O3	O4–O6
Someone else already reported it.	0.39% (0.00–5.88)	1.02% (0.00–12.15)	NR (0.00–31.46)	NR (NR)	NR (NR)
You thought it was not serious enough to report.	8.75% (3.17–18.41)	NR (1.71–26.88)	NR (0.05–21.59)	NR (NR)	NR (NR)
You did not want more people to know.	3.11% (0.43–10.32)	5.16% (0.48–18.88)	NR (0.00–31.46)	NR (NR)	NR (NR)
You wanted to forget about it and move on.	NR (6.98–32.78)	NR (8.06–44.13)	NR (1.01–45.90)	NR (NR)	NR (NR)
You did not think your report would be kept confidential.	3.20% (0.36–11.41)	NR (0.15–23.43)	NR (0.02–20.80)	NR (NR)	NR (NR)
You did not think anything would be done.	1.46% (0.04–7.94)	NR (0.11–19.04)	NR (0.00–31.46)	NR (NR)	NR (NR)
You did not think you would be believed.	2.63% (0.28–9.59)	2.17% (0.02–14.10)	NR (0.00–31.46)	NR (NR)	NR (NR)
You did not trust the process would be fair.	3.47% (0.37–12.48)	0.00% (0.00–12.06)	NR (0.15–24.15)	NR (NR)	NR (NR)
You felt partially to blame.	5.99% (1.50–15.33)	6.38% (0.77–21.34)	NR (0.91–28.25)	NR (NR)	NR (NR)
You thought other people would blame you.	1.49% (0.04–8.11)	0.00% (0.00–12.06)	NR (0.03–21.28)	NR (NR)	NR (NR)
You thought you might get in trouble for something you did (for example, underage drinking or fraternization).	1.82% (0.09–8.35)	2.39% (0.03–14.45)	NR (0.00–17.91)	NR (NR)	NR (NR)
You thought you might be labeled as a troublemaker.	0.88% (0.00–6.77)	0.00% (0.00–12.06)	NR (0.00–19.01)	NR (NR)	NR (NR)
You thought it might hurt your career.	NR% (0.49–41.09)	0.00% (0.00–12.06)	NR (1.60–69.82)	NR (NR)	NR (NR)
You did not want to hurt the person's career or family.	4.17% (0.75–12.44)	4.72% (0.35–18.50)	NR (0.26–24.39)	NR (NR)	NR (NR)
You were worried about retaliation by the person(s) who did it.	2.01% (0.13–8.65)	5.32% (0.48–19.65)	NR (0.00–31.46)	NR (NR)	NR (NR)

Table A.33.c—Continued

	Total	E1–E4	E5–E9	O1–O3	O4–O6
You were worried about retaliation by supervisor or someone in your chain of command.	1.42% (0.03–7.74)	0.00% (0.00–12.06)	NR (0.03–21.02)	NR (NR)	NR (NR)
You took other actions to handle the situation.	NR (15.87–51.28)	NR (14.93–53.81)	NR (5.86–60.21)	NR (NR)	NR (NR)

NOTE: Includes estimates for active-component Coast Guard members. 95-percent confidence intervals for each estimate are included in parentheses.

Some response options that appear in Table A.33.b do not appear here because they were not selected by any Coast Guard respondent.

NR = Not reportable.

A.34. Percentage of members who would make the same reporting decision again among members who experienced a sexual assault in the past year

Table A.34.a
Percentage of members who would make the same reporting decision again among members who experienced a sexual assault in the past year, by gender

SAFU32: In retrospect, would you make the same decision about reporting if you could do it over?

Total	Men	Women
82.46%	NR	76.26%
(70.39–91.11)	(NR)	(63.15–86.52)

NOTE: Includes estimates for active-component Coast Guard members. 95-percent confidence intervals for each estimate are included in parentheses.

NR = Not reportable.

Table A.34.b
Percentage of members who would make the same reporting decision again among members who experienced a sexual assault in the past year, by service

SAFU32: In retrospect, would you make the same decision about reporting if you could do it over?

Total DoD	Army	Navy	Air Force	Marine Corps	Coast Guard
72.00%	72.77%	68.60%	79.67%	72.08%	82.46%
(66.76–76.84)	(65.76–79.04)	(56.94–78.75)	(75.32–83.56)	(58.61–83.15)	(70.39–91.11)

NOTE: Includes estimates for active-component DoD and Coast Guard members. 95-percent confidence intervals for each estimate are included in parentheses.

Table A.34.c
Percentage of members who would make the same reporting decision again among members who experienced a sexual assault in the past year, by pay grade

SAFU32: In retrospect, would you make the same decision about reporting if you could do it over?

Total	E1–E4	E5–E9	O1–O3	O4–O6
82.46%	NR	NR	NR	NR
(70.39–91.11)	(53.26–85.36)	(69.06–97.27)	(NR)	(NR)

NOTE: Includes estimates for active-component Coast Guard members. 95-percent confidence intervals for each estimate are included in parentheses.

Too few warrant officers were included in the sample to break them out as a separate pay grade. For the purposes of this table, warrant officers have been included with the E5–E9 category.

NR = Not reportable.

A.35. Rates of perceived retaliation or negative career actions following the single or most serious assault among members who experienced a sexual assault in the past year

Table A.35.a
Rates of perceived retaliation or negative career actions following the single or most serious assault among members who experienced a sexual assault in the past year, by gender

SAFU33: As a result of the unwanted event, did you...

	Total	Men	Women
Experience any professional retaliation? For example, loss of privileges, denied promotion/training, transferred to a less favorable job.			
Yes	NR (2.46–52.84)	NR (NR)	7.52% (2.08–18.17)
No	NR (46.56–94.74)	NR (NR)	87.74% (76.26–94.96)
Do not know	4.31% (1.07–11.21)	NR (NR)	4.74% (0.99–13.21)
Experience any social retaliation? For example, ignored by coworkers, being blamed for what happened.			
Yes	NR (13.15–57.89)	NR (NR)	29.35% (17.37–43.85)
No	NR (40.26–83.91)	NR (NR)	67.82% (53.52–80.04)
Do not know	3.11% (0.57–9.31)	NR (NR)	2.83% (0.39–9.42)
Experience any administrative actions that you did not want? For example, placed on a legal or medical hold, denied a deployment opportunity, transferred to a different assignment.			
Yes	NR (1.30–53.17)	NR (NR)	3.71% (0.52–12.12)
No	NR (46.73–95.45)	NR (NR)	89.27% (77.32–96.22)
Do not know	5.76% (1.52–14.38)	NR (NR)	7.03% (1.61–18.50)

Table A.35.a—Continued

	Total	Men	Women
Experience any punishments for infractions/violations such as underage drinking or fraternization?			
Yes	NR (0.81–53.70)	NR (NR)	1.57% (0.04–8.44)
No	NR (46.91–98.58)	NR (NR)	97.94% (91.67–99.84)
Do not know	1.63% (0.10–7.12)	NR (NR)	0.49% (0.00–5.62)

NOTE: Includes estimates for active-component Coast Guard members. 95-percent confidence intervals for each estimate are included in parentheses.

NR = Not reportable.

Table A.35.b
Rates of perceived retaliation or negative career actions following the single or most serious assault among members who experienced a sexual assault in the past year, by service

SAFU33: As a result of the unwanted event, did you...

	Total DoD	Army	Navy	Air Force	Marine Corps	Coast Guard
Experience any professional retaliation? For example, loss of privileges, denied promotion/training, transferred to a less favorable job.						
Yes	14.31% (9.96–19.65)	13.65% (9.95–18.09)	14.98% (5.69–29.82)	13.88% (9.42–19.44)	14.60% (5.79–28.50)	NR (2.46–52.84)
No	76.54% (71.06–81.44)	70.84% (62.97–77.89)	78.74% (65.69–88.56)	80.80% (74.80–85.90)	81.08% (67.46–90.79)	NR (46.56–94.74)
Do not know	9.15% (6.43–12.54)	15.51% (9.14–23.95)	6.28% (3.74–9.80)	5.32% (2.68–9.35)	4.32% (1.94–8.19)	4.31% (1.07–11.21)
Experience any social retaliation? For example, ignored by coworkers, being blamed for what happened.						
Yes	26.86% (21.39–32.90)	24.55% (18.34–31.66)	29.84% (18.24–43.71)	23.48% (18.60–28.94)	NR (14.57–45.33)	NR (13.15–57.89)
No	65.95% (60.04–71.51)	64.77% (56.53–72.40)	65.12% (52.32–76.51)	71.43% (65.60–76.79)	NR (49.15–80.51)	NR (40.26–83.91)
Do not know	7.20% (4.80–10.29)	10.68% (5.20–18.85)	5.04% (2.99–7.90)	5.09% (2.47–9.15)	5.78% (2.46–11.24)	3.11% (0.57–9.31)
Experience any administrative actions that you did not want? For example, placed on a legal or medical hold, denied a deployment opportunity, transferred to a different assignment.						
Yes	9.02% (6.58–11.99)	12.24% (7.56–18.42)	4.53% (2.58–7.30)	10.20% (6.41–15.19)	11.08% (3.47–24.71)	NR (1.30–53.17)
No	84.16% (80.13–87.66)	75.81% (67.28–83.06)	91.41% (87.13–94.63)	85.80% (80.29–90.24)	85.47% (72.39–93.93)	NR (46.73–95.45)
Do not know	6.82% (4.31–10.16)	11.95% (6.10–20.41)	4.06% (1.82–7.71)	4.01% (1.74–7.76)	3.45% (1.35–7.13)	5.76% (1.52–14.38)
Experience any punishments for infractions/violations such as underage drinking or fraternization?						
Yes	4.04% (2.15–6.84)	6.18% (2.30–12.97)	1.48% (0.54–3.20)	1.93% (0.88–3.64)	6.96% (1.34–19.63)	NR (0.81–53.70)
No	88.89% (84.09–92.65)	84.33% (75.26–91.07)	94.74% (90.95–97.27)	96.02% (93.72–97.66)	NR (56.72–93.50)	NR (46.91–98.58)
Do not know	7.07% (3.89–11.65)	9.49% (4.15–17.96)	3.78% (1.60–7.45)	2.05% (0.91–3.93)	NR (2.07–39.16)	1.63% (0.10–7.12)

NOTE: Includes estimates for active-component DoD and Coast Guard members. 95-percent confidence intervals for each estimate are included in parentheses.

NR = Not reportable.

Table A.35.c
Rates of perceived retaliation or negative career actions following the single or most serious assault among members who experienced a sexual assault in the past year, by pay grade

SAFU33: As a result of the unwanted event, did you...

	Total	E1–E4	E5–E9	O1–O3	O4–O6
Experience any professional retaliation? For example, loss of privileges, denied promotion/training, transferred to a less favorable job.					
Yes	NR (2.46–52.84)	NR (3.05–29.04)	NR (2.34–78.72)	NR (NR)	NR (NR)
No	NR (46.56–94.74)	NR (69.63–95.65)	NR (21.20–95.18)	NR (NR)	NR (NR)
Do not know	4.31% (1.07–11.21)	1.89% (0.03–11.40)	NR (0.22–21.01)	NR (NR)	NR (NR)
Experience any social retaliation? For example, ignored by coworkers, being blamed for what happened.					
Yes	NR (13.15–57.89)	NR (24.41–60.15)	NR (4.23–78.73)	NR (NR)	NR (NR)
No	NR (40.26–83.91)	NR (37.26–72.75)	NR (20.70–93.11)	NR (NR)	NR (NR)
Do not know	3.11% (0.57–9.31)	3.02% (0.17–13.19)	NR (0.13–20.11)	NR (NR)	NR (NR)
Experience any administrative actions that you did not want? For example, placed on a legal or medical hold, denied a deployment opportunity, transferred to a different assignment.					
Yes	NR (1.30–53.17)	4.71% (0.32–18.97)	NR (2.52–78.69)	NR (NR)	NR (NR)
No	NR (46.73–95.45)	NR (71.46–97.87)	NR (21.15–94.92)	NR (NR)	NR (NR)
Do not know	5.76% (1.52–14.38)	NR (0.38–23.68)	NR (0.22–21.01)	NR (NR)	NR (NR)
Experience any punishments for infractions/violations such as underage drinking or fraternization?					
Yes	NR (0.81–53.70)	0.00% (0.00–9.19)	NR (2.85–78.66)	NR (NR)	NR (NR)
No	NR (46.91–98.58)	99.14% (90.36–100.00)	NR (21.26–95.72)	NR (NR)	NR (NR)
Do not know	1.63% (0.10–7.12)	0.86% (0.00–9.64)	NR (0.02–17.90)	NR (NR)	NR (NR)

NOTE: Includes estimates for active-component Coast Guard members. 95-percent confidence intervals for each estimate are included in parentheses.

Too few warrant officers were included in the sample to break them out as a separate pay grade. For the purposes of this table, warrant officers have been included with the E5–E9 category.

NR = Not reportable.

A.36. Sources of perceived social retaliation among members who indicated social retaliation following the single or most serious sexual assault in the past year

Table A.36.a
Sources of perceived social retaliation among members who indicated social retaliation following the single or most serious sexual assault in the past year, by gender

SAFU33b: Who socially retaliated against you?

	Total	Men	Women
Someone who outranks you	NR (45.71–96.20)	NR (NR)	NR (33.53–85.31)
Someone who is a similar rank or below you in rank	NR (43.73–95.71)	NR (NR)	NR (31.51–83.28)
Non-military personnel	NR (14.39–91.66)	NR (NR)	NR (4.68–52.23)
Do not know who they were	NR (4.04–91.19)	NR (NR)	NR (0.00–21.21)

NOTE: Includes estimates for active-component Coast Guard members. 95-percent confidence intervals for each estimate are included in parentheses.

NR = Not reportable.

Table A.36.b
Sources of perceived social retaliation among members who indicated social retaliation following the single or most serious sexual assault in the past year, by service

SAFU33b: Who socially retaliated against you?

	Total DoD	Army	Navy	Air Force	Marine Corps	Coast Guard
Someone who outranks you	66.30% (54.91–76.44)	NR (49.10–78.24)	NR (46.83–86.15)	62.20% (50.54–72.91)	NR (33.66–89.78)	NR (45.71–96.20)
Someone who is a similar rank or below you in rank	67.39% (52.74–79.92)	71.33% (58.09–82.34)	NR (33.47–90.37)	66.54% (52.70–78.57)	NR (28.97–88.41)	NR (43.73–95.71)
Non-military personnel	10.60% (5.88–17.22)	11.63% (6.59–18.59)	5.72% (1.60–13.94)	14.37% (7.52–24.02)	NR (1.89–56.84)	NR (14.39–91.66)
Do not know who they were	8.98% (3.02–19.61)	NR (1.90–39.40)	2.56% (0.49–7.59)	3.99% (0.76–11.62)	NR (2.65–57.29)	NR (4.04–91.19)

NOTE: Includes estimates for active-component DoD and Coast Guard members. 95-percent confidence intervals for each estimate are included in parentheses.

NR = Not reportable.

Table A.36.c
Sources of perceived social retaliation among members who indicated social retaliation following the single or most serious sexual assault in the past year, by pay grade

SAFU33b: Who socially retaliated against you?

	Total	E1–E4	E5–E9	O1–O3	O4–O6
Someone who outranks you	NR (45.71–96.20)	NR (NR)	NR (NR)	NR (NR)	NR (NR)
Someone who is a similar rank or below you in rank	NR (43.73–95.71)	NR (NR)	NR (NR)	NR (NR)	NR (NR)
Non-military personnel	NR (14.39–91.66)	NR (NR)	NR (NR)	NR (NR)	NR (NR)
Do not know who they were	NR (4.04–91.19)	NR (NR)	NR (NR)	NR (NR)	NR (NR)

NOTE: Includes estimates for active-component Coast Guard members. 95-percent confidence intervals for each estimate are included in parentheses.

Too few warrant officers were included in the sample to break them out as a separate pay grade. For the purposes of this table, warrant officers have been included with the E5–E9 category.

NR = Not reportable.

A.37. Members who completed a Victim Reporting Preference Statement for any sexual assault among members who experienced a sexual assault in the past year

Table A.37.a

Members who completed a Victim Reporting Preference Statement for any sexual assault among members who experienced a sexual assault in the past year, by gender

SAFU36: Did you initial and sign a form labeled VICTIM REPORTING PREFERENCE STATEMENT (DD Form 2910 or CG Form 6095)? This form allows you to decide whether to make a restricted or unrestricted report of sexual assault.[1]

	Total	Men	Women
Yes	NR	NR	15.02%
	(6.60–53.94)	(NR)	(7.11–26.60)
No	NR	NR	78.36%
	(44.26–89.92)	(NR)	(65.86–87.94)
Not sure	4.17%	NR	6.61%
	(1.04–10.82)	(NR)	(1.97–15.53)

NOTE: Includes estimates for active-component Coast Guard members. 95-percent confidence intervals for each estimate are included in parentheses.

NR = Not reportable.

[1] Question continues: "A Sexual Assault Response Coordinator (SARC) or Victim Advocate (VA) would have assisted you with completing this form. To see a version of this form, click here [hyperlink to image of DD Form 2910]." Question was preceded by, "Thank you for sharing these details about the unwanted event you chose as the worst or most serious. For the next question, please consider any unwanted event that happened to you."

Table A.37.b

Members who completed a Victim Reporting Preference Statement for any sexual assault among members who experienced a sexual assault in the past year, by service

SAFU36: Did you initial and sign a form labeled VICTIM REPORTING PREFERENCE STATEMENT (DD Form 2910 or CG Form 6095)? This form allows you to decide whether to make a restricted or unrestricted report of sexual assault.

	Total DoD	Army	Navy	Air Force	Marine Corps	Coast Guard
Yes	11.24% (9.21–13.54)	12.96% (9.68–16.86)	8.20% (5.53–11.63)	13.17% (10.42–16.33)	12.85% (5.70–23.78)	NR (6.60–53.94)
No	78.08% (74.31–81.53)	73.52% (66.14–80.04)	83.83% (78.62–88.20)	75.46% (70.91–79.61)	77.38% (64.67–87.23)	NR (44.26–89.92)
Not sure	10.68% (8.00–13.88)	13.52% (7.78–21.29)	7.97% (5.09–11.75)	11.38% (8.32–15.07)	9.77% (4.56–17.77)	4.17% (1.04–10.82)

NOTE: Includes estimates for active-component DoD and Coast Guard members. 95-percent confidence intervals for each estimate are included in parentheses.

NR = Not reportable.

Table A.37.c

Members who completed a Victim Reporting Preference Statement for any sexual assault among members who experienced a sexual assault in the past year, by pay grade

SAFU36: Did you initial and sign a form labeled VICTIM REPORTING PREFERENCE STATEMENT (DD Form 2910 or CG Form 6095)? This form allows you to decide whether to make a restricted or unrestricted report of sexual assault.

	Total	E1–E4	E5–E9	O1–O3	O4–O6
Yes	NR (6.60–53.94)	NR (8.45–37.02)	NR (4.80–78.81)	NR (NR)	NR (NR)
No	NR (44.26–89.92)	NR (54.10–85.63)	NR (21.14–94.86)	NR (NR)	NR (NR)
Not sure	4.17% (1.04–10.82)	8.17% (1.74–21.92)	0.54% (0.00–14.22)	NR (NR)	NR (NR)

NOTE: Includes estimates for active-component Coast Guard members. 95-percent confidence intervals for each estimate are included in parentheses.

Too few warrant officers were included in the sample to break them out as a separate pay grade. For the purposes of this table, warrant officers have been included with the E5–E9 category.

NR = Not reportable.

A.38. Types of services offered to members who experienced a sexual assault in the past year and completed a Victim Reporting Preference Statement

Table A.38.a

Types of services offered to members who experienced a sexual assault in the past year and completed a Victim Reporting Preference Statement, by gender

SAFU37: When you reported the event, were you offered...

	Total	Men	Women
Sexual assault advocacy services (e.g., referrals or offers to accompany/ transport you to appointments)?	NR (79.56–100.00)	NR (NR)	NR (74.27–99.95)
Counseling services?	NR (79.56–100.00)	NR (NR)	NR (74.27–99.95)
Medical or forensic services?	NR (52.59–97.95)	NR (NR)	NR (35.01–88.50)
Legal services?	NR (71.77–99.48)	NR (NR)	NR (57.22–97.37)
Chaplain services?	NR (73.02–99.66)	NR (NR)	NR (58.21–98.29)

NOTE: Includes estimates for active-component Coast Guard members. 95-percent confidence intervals for each estimate are included in parentheses.

NR = Not reportable.

Table A.38.b
Types of services offered to members who experienced a sexual assault in the past year and completed a Victim Reporting Preference Statement, by service

SAFU37: When you reported the event, were you offered...

	Total DoD	Army	Navy	Air Force	Marine Corps	Coast Guard
Sexual assault advocacy services (e.g., referrals or offers to accompany/ transport you to appointments)?	59.65% (49.80–68.96)	NR (38.08–73.31)	57.91% (44.01–70.95)	75.78% (67.25–83.02)	NR (34.93–79.36)	NR (79.56–100.00)
Counseling services?	66.82% (58.10–74.76)	NR (55.94–85.14)	57.23% (43.36–70.33)	79.14% (70.69–86.07)	NR (35.79–75.51)	NR (79.56–100.00)
Medical or forensic services?	55.69% (46.28–64.81)	NR (36.43–70.32)	53.63% (40.05–66.83)	64.64% (55.50–73.05)	NR (33.94–78.57)	NR (52.59–97.95)
Legal services?	60.56% (51.94–68.72)	NR (48.32–78.24)	52.42% (39.03–65.55)	75.62% (67.47–82.60)	NR (31.46–68.75)	NR (71.77–99.48)
Chaplain services?	61.23% (51.59–70.27)	NR (43.31–79.70)	54.79% (41.18–67.90)	71.28% (62.25–79.22)	NR (36.05–80.31)	NR (73.02–99.66)

NOTE: Includes estimates for active-component DoD and Coast Guard members. 95-percent confidence intervals for each estimate are included in parentheses.

NR = Not reportable.

Table A.38.c
Types of services offered to members who experienced a sexual assault in the past year and completed a Victim Reporting Preference Statement, by pay grade

SAFU37: When you reported the event, were you offered...

	Total	E1–E4	E5–E9	O1–O3	O4–O6
Sexual assault advocacy services (e.g., referrals or offers to accompany/transport you to appointments)?	NR (79.56–100.00)	NR (NR)	NR (NR)	NR (NR)	NR (NR)
Counseling services?	NR (79.56–100.00)	NR (NR)	NR (NR)	NR (NR)	NR (NR)
Medical or forensic services?	NR (52.59–97.95)	NR (NR)	NR (NR)	NR (NR)	NR (NR)
Legal services?	NR (71.77–99.48)	NR (NR)	NR (NR)	NR (NR)	NR (NR)
Chaplain services?	NR (73.02–99.66)	NR (NR)	NR (NR)	NR (NR)	NR (NR)

NOTE: Includes estimates for active-component Coast Guard members. 95-percent confidence intervals for each estimate are included in parentheses.

Too few warrant officers were included in the sample to break them out as a separate pay grade. For the purposes of this table, warrant officers have been included with the E5–E9 category.

NR = Not reportable.

Sexual Harassment and Gender Discrimination in the Coast Guard: Detailed Results

B.1. Percentage of members who experienced a sexually hostile work environment in the past year

Table B.1
Percentage of members who experienced a sexually hostile work environment in the past year, by gender and pay grade

Pay Grade	Total	Men	Women
E1–E4	8.63% (6.97–10.54)	5.07% (3.34–7.34)	23.01% (19.48–26.86)
E5–E9	4.55% (3.58–5.69)	3.30% (2.31–4.55)	15.75% (12.54–19.40)
O1–O3	7.44% (5.27–10.14)	3.79% (1.70–7.20)	19.58% (14.74–25.19)
O4–O6	2.74% (1.38–4.86)	1.47% (0.31–4.20)	11.06% (6.35–17.52)

NOTE: Includes estimates for active-component Coast Guard members. 95-percent confidence intervals for each estimate are included in parentheses.

Too few warrant officers were included in the sample to break them out as a separate pay grade. For the purposes of this table, warrant officers have been included with the E5–E9 category.

B.2. Percentage of members who experienced sexual quid pro quo in the past year

Table B.2
Percentage of members who experienced sexual quid pro quo in the past year, by gender and pay grade

Pay Grade	Total	Men	Women
E1–E4	0.11% (0.01–0.44)	0.00% (0.00–0.70)	0.56% (0.14–1.48)
E5–E9	0.06% (0.00–0.25)	0.00% (0.00–0.26)	0.56% (0.16–1.40)
O1–O3	0.00% (0.00–0.96)	0.00% (0.00–1.52)	0.00% (0.00–1.13)
O4–O6	0.13% (0.00–1.00)	0.00% (0.00–1.44)	0.94% (0.03–5.08)

NOTE: Includes estimates for active-component Coast Guard members. 95-percent confidence intervals for each estimate are included in parentheses.

Too few warrant officers were included in the sample to break them out as a separate pay grade. For the purposes of this table, warrant officers have been included with the E5–E9 category.

B.3. Percentage of members who experienced sexual harassment in the past year

Table B.3
Percentage of members who experienced sexual harassment in the past year, by gender and pay grade

Pay Grade	Total	Men	Women
E1–E4	8.71% (7.03–10.63)	5.11% (3.37–7.39)	23.11% (19.56–26.96)
E5–E9	4.56% (3.58–5.70)	3.30% (2.32–4.55)	15.76% (12.55–19.41)
O1–O3	7.44% (5.27–10.14)	3.79% (1.70–7.20)	19.60% (14.76–25.22)
O4–O6	2.74% (1.38–4.86)	1.47% (0.31–4.20)	11.06% (6.35–17.52)

NOTE: Includes estimates for active-component Coast Guard members. 95-percent confidence intervals for each estimate are included in parentheses.

Too few warrant officers were included in the sample to break them out as a separate pay grade. For the purposes of this table, warrant officers have been included with the E5–E9 category.

B.4. Percentage of members who experienced gender discrimination in the past year

Table B.4
Percentage of members who experienced gender discrimination in the past year, by gender and pay grade

Pay Grade	Total	Men	Women
E1–E4	2.79% (2.02–3.74)	0.76% (0.21–1.94)	10.93% (8.41–13.90)
E5–E9	2.08% (1.37–3.03)	1.15% (0.48–2.31)	10.46% (7.94–13.44)
O1–O3	4.86% (3.19–7.05)	1.93% (0.52–4.90)	14.62% (10.48–19.63)
O4–O6	2.57% (1.35–4.41)	0.35% (0.01–1.91)	17.08% (11.05–24.67)

NOTE: Includes estimates for active-component Coast Guard members. 95-percent confidence intervals for each estimate are included in parentheses.

Too few warrant officers were included in the sample to break them out as a separate pay grade. For the purposes of this table, warrant officers have been included with the E5–E9 category.

B.5. Percentage of members who experienced sexual harassment or gender discrimination in the past year

Table B.5
Percentage of members who experienced sexual harassment or gender discrimination in the past year, by gender and pay grade

Pay Grade	Total	Men	Women
E1–E4	9.59%	5.57%	25.74%
	(7.86–11.56)	(3.77–7.89)	(22.09–29.65)
E5–E9	5.80%	4.18%	20.31%
	(4.63–7.17)	(2.98–5.69)	(16.80–24.20)
O1–O3	9.55%	5.18%	24.10%
	(7.04–12.59)	(2.65–9.00)	(18.93–29.91)
O4–O6	4.47%	1.82%	21.82%
	(2.81–6.71)	(0.53–4.45)	(15.24–29.65)

NOTE: Includes estimates for active-component Coast Guard members. 95-percent confidence intervals for each estimate are included in parentheses.

Too few warrant officers were included in the sample to break them out as a separate pay grade. For the purposes of this table, warrant officers have been included with the E5–E9 category.

B.6. Percentage of members who experienced each type of sexual harassment or gender discrimination violation in the past year

Table B.6.a
Percentage of members who experienced each type of sexual harassment or gender discrimination violation in the past year, by gender

SH1-SH15: In this section, you will be asked about several things that someone from work might have done to you that were upsetting or offensive. Since [X date], ...

	Total	Men	Women
Did someone from work repeatedly tell sexual 'jokes' that made you uncomfortable, angry, or upset? (Follow-up questions established that actions were also persistent or severe.)	2.76% (2.25–3.34)	1.48% (0.99–2.13)	10.17% (8.54–11.99)
Did someone from work embarrass, anger, or upset you by repeatedly suggesting that you do not act like a [man/woman] is supposed to? (Follow-up questions established that actions were also persistent or severe.)	1.94% (1.51–2.46)	1.34% (0.89–1.93)	5.46% (4.27–6.86)
Did someone from work repeatedly make sexual gestures or sexual body movements (for example, thrusting their pelvis or grabbing their crotch) that made you uncomfortable, angry, or upset? (Follow-up questions established that actions were also persistent or severe.)	1.12% (0.79–1.53)	0.72% (0.40–1.19)	3.45% (2.43–4.73)
Did someone from work display, show, or send sexually explicit materials like pictures or videos that made you uncomfortable, angry, or upset? (Follow-up questions established that actions were also persistent or severe.)	0.79% (0.56–1.08)	0.42% (0.21–0.75)	2.92% (2.07–4.00)
Did someone from work repeatedly tell you about their sexual activities in a way that made you uncomfortable, angry, or upset? (Follow-up questions established that actions were also persistent or severe.)	1.37% (1.02–1.81)	0.76% (0.43–1.25)	4.93% (3.73–6.38)
Did someone from work repeatedly ask you questions about your sex life or sexual interests that made you uncomfortable, angry, or upset? (Follow-up questions established that actions were also persistent or severe.)	1.16% (0.86–1.52)	0.59% (0.33–0.98)	4.43% (3.32–5.79)
Did someone from work make repeated sexual comments about your appearance or body that made you uncomfortable, angry, or upset? (Follow-up questions established that actions were also persistent or severe.)	1.24% (0.92–1.64)	0.46% (0.19–0.93)	5.76% (4.49–7.25)

Table B.6.a—Continued

	Total	Men	Women
Did someone from work either take or share sexually suggestive pictures or videos of you when you did not want them to? (Follow-up questions established that actions were also persistent or severe.)	0.16% (0.06–0.34)	0.10% (0.01–0.33)	0.53% (0.23–1.05)
Did someone from work make repeated attempts to establish an unwanted romantic or sexual relationship with you? (Follow-up questions established that actions were also persistent or severe.)	0.72% (0.51–0.99)	0.19% (0.05–0.50)	3.83% (2.86–5.01)
Did someone from work intentionally touch you in a sexual way when you did not want them to? (Categorized as severe without requiring additional follow-up questions)	0.39% (0.24–0.59)	0.14% (0.04–0.35)	1.83% (1.09–2.87)
Did someone from work repeatedly touch you in any other way that made you uncomfortable, angry, or upset? (Follow-up questions established that actions were also persistent or severe.)	1.47% (1.09–1.94)	0.75% (0.40–1.28)	5.65% (4.35–7.19)
Has someone from work made you feel as if you would get some workplace benefit in exchange for doing something sexual? (Follow-up questions established that target had direct evidence of an exchange.)	0.05% (0.01–0.16)	0.00% (0.00–0.16)	0.37% (0.16–0.72)
Has someone from work made you feel like you would get punished or treated unfairly in the workplace if you did not do something sexual? (Follow-up questions established that target had direct evidence of an exchange.)	0.04% (0.00–0.14)	0.00% (0.00–0.16)	0.25% (0.07–0.64)
Did you hear someone from work say that [men/women] are not as good as [women/men] at your particular job, or that [men/women] should be prevented from having your job? (Follow-up question established that action resulted in harm to the target's career)	1.33% (0.95–1.81)	0.39% (0.09–1.08)	6.80% (5.52–8.27)
Do you think someone from work mistreated, ignored, excluded, or insulted you because you are a [man/woman]? (Follow-up question established that action resulted in harm to the target's career)	2.47% (1.98–3.04)	1.03% (0.57–1.71)	10.81% (9.23–12.54)

NOTE: Includes estimates for active-component Coast Guard members. 95-percent confidence intervals for each estimate are included in parentheses.

Table B.6.b
Percentage of members who experienced each type of sexual harassment or gender discrimination violation in the past year, by service

SH1–SH15: In this section, you will be asked about several things that someone from work might have done to you that were upsetting or offensive. Since [X date], ...

	Total DoD	Army	Navy	Air Force	Marine Corps	Coast Guard
Did someone from work repeatedly tell sexual 'jokes' that made you uncomfortable, angry, or upset? (Follow-up questions established that actions were also persistent or severe.)	3.79% (3.49–4.10)	4.19% (3.71–4.71)	5.25% (4.47–6.12)	2.30% (2.02–2.62)	2.72% (1.99–3.62)	2.76% (2.25–3.34)
Did someone from work embarrass, anger, or upset you by repeatedly suggesting that you do not act like a [man/woman] is supposed to? (Follow-up questions established that actions were also persistent or severe.)	3.98% (3.62–4.37)	4.66% (4.04–5.36)	5.01% (4.17–5.95)	1.70% (1.41–2.02)	4.26% (3.15–5.61)	1.94% (1.51–2.46)
Did someone from work repeatedly make sexual gestures or sexual body movements (for example, thrusting their pelvis or grabbing their crotch) that made you uncomfortable, angry, or upset? (Follow-up questions established that actions were also persistent or severe.)	1.97% (1.73–2.24)	2.24% (1.84–2.69)	2.98% (2.29–3.81)	0.76% (0.64–0.90)	1.57% (1.02–2.30)	1.12% (0.79–1.53)
Did someone from work display, show, or send sexually explicit materials like pictures or videos that made you uncomfortable, angry, or upset? (Follow-up questions established that actions were also persistent or severe.)	1.13% (0.97–1.31)	1.31% (1.02–1.67)	1.51% (1.12–1.99)	0.53% (0.44–0.63)	0.98% (0.65–1.43)	0.79% (0.56–1.08)
Did someone from work repeatedly tell you about their sexual activities in a way that made you uncomfortable, angry, or upset? (Follow-up questions established that actions were also persistent or severe.)	2.42% (2.20–2.65)	2.74% (2.35–3.17)	3.41% (2.85–4.05)	1.29% (1.14–1.46)	1.78% (1.32–2.34)	1.37% (1.02–1.81)

Table B.6.b—Continued

	Total DoD	Army	Navy	Air Force	Marine Corps	Coast Guard
Did someone from work repeatedly ask you questions about your sex life or sexual interests that made you uncomfortable, angry, or upset? (Follow-up questions established that actions were also persistent or severe.)	2.29% (2.05–2.54)	2.67% (2.27–3.11)	3.48% (2.81–4.26)	1.00% (0.87–1.14)	1.40% (1.04–1.84)	1.16% (0.86–1.52)
Did someone from work make repeated sexual comments about your appearance or body that made you uncomfortable, angry, or upset? (Follow-up questions established that actions were also persistent or severe.)	2.23% (1.99–2.49)	2.32% (1.99–2.68)	3.28% (2.66–4.01)	0.94% (0.82–1.07)	2.42% (1.56–3.57)	1.24% (0.92–1.64)
Did someone from work either take or share sexually suggestive pictures or videos of you when you did not want them to? (Follow-up questions established that actions were also persistent or severe.)	0.34% (0.25–0.45)	0.50% (0.30–0.76)	0.27% (0.18–0.38)	0.13% (0.09–0.19)	0.38% (0.15–0.79)	0.16% (0.06–0.34)
Did someone from work make repeated attempts to establish an unwanted romantic or sexual relationship with you? (Follow-up questions established that actions were also persistent or severe.)	1.47% (1.33–1.63)	1.58% (1.41–1.77)	2.27% (1.79–2.84)	0.72% (0.62–0.84)	1.11% (0.81–1.47)	0.72% (0.51–0.99)
Did someone from work intentionally touch you in a sexual way when you did not want them to? (Categorized as severe without requiring additional follow-up questions)	1.46% (1.26–1.69)	1.42% (1.13–1.76)	2.26% (1.68–2.97)	0.60% (0.48–0.74)	1.70% (1.07–2.55)	0.39% (0.24–0.59)
Did someone from work repeatedly touch you in any other way that made you uncomfortable, angry, or upset? (Follow-up questions established that actions were also persistent or severe.)	2.70% (2.44–2.99)	2.79% (2.41–3.20)	4.09% (3.31–5.00)	1.31% (1.10–1.55)	2.47% (1.79–3.32)	1.47% (1.09–1.94)
Has someone from work made you feel as if you would get some workplace benefit in exchange for doing something sexual? (Follow-up questions established that target had direct evidence of an exchange.)	0.47% (0.35–0.63)	0.55% (0.40–0.74)	0.72% (0.36–1.29)	0.12% (0.08–0.17)	0.44% (0.11–1.17)	0.05% (0.01–0.16)

Table B.6.b—Continued

	Total DoD	Army	Navy	Air Force	Marine Corps	Coast Guard
Has someone from work made you feel like you would get punished or treated unfairly in the workplace if you did not do something sexual? (Follow-up questions established that target had direct evidence of an exchange.)	0.30% (0.22–0.40)	0.40% (0.28–0.56)	0.27% (0.16–0.43)	0.08% (0.05–0.13)	0.42% (0.10–1.15)	0.04% (0.00–0.14)
Did you hear someone from work say that [men/women] are not as good as [women/men] at your particular job, or that [men/women] should be prevented from having your job? (Follow-up question established that action resulted in harm to the target's career)	1.76% (1.64–1.88)	2.05% (1.83–2.30)	2.47% (2.18–2.79)	0.93% (0.81–1.05)	1.14% (0.88–1.47)	1.33% (0.95–1.81)
Do you think someone from work mistreated, ignored, excluded, or insulted you because you are a [man/woman]? (Follow-up question established that action resulted in harm to the target's career)	2.91% (2.72–3.11)	3.35% (3.05–3.67)	4.09% (3.54–4.71)	1.72% (1.56–1.89)	1.72% (1.39–2.10)	2.47% (1.98–3.04)

NOTE: Includes estimates for active-component DoD and Coast Guard members. 95-percent confidence intervals for each estimate are included in parentheses.

Table B.6.c
Percentage of members who experienced each type of sexual harassment or gender discrimination violation in the past year, by pay grade

SH1-SH15: In this section, you will be asked about several things that someone from work might have done to you that were upsetting or offensive. Since [X date], ...

	Total	E1–E4	E5–E9	O1–O3	O4–O6
Did someone from work repeatedly tell sexual 'jokes' that made you uncomfortable, angry, or upset? (Follow-up questions established that actions were also persistent or severe.)	2.76% (2.25–3.34)	4.22% (3.11–5.58)	2.08% (1.45–2.88)	2.51% (1.45–4.03)	1.49% (0.55–3.22)
Did someone from work embarrass, anger, or upset you by repeatedly suggesting that you do not act like a [man/woman] is supposed to? (Follow-up questions established that actions were also persistent or severe.)	1.94% (1.51–2.46)	2.87% (1.94–4.08)	1.50% (0.97–2.22)	1.89% (0.95–3.35)	1.09% (0.30–2.76)
Did someone from work repeatedly make sexual gestures or sexual body movements (for example, thrusting their pelvis or grabbing their crotch) that made you uncomfortable, angry, or upset? (Follow-up questions established that actions were also persistent or severe.)	1.12% (0.79–1.53)	1.36% (0.83–2.09)	1.25% (0.73–1.98)	0.32% (0.05–1.06)	0.25% (0.01–1.21)
Did someone from work display, show, or send sexually explicit materials like pictures or videos that made you uncomfortable, angry, or upset? (Follow-up questions established that actions were also persistent or severe.)	0.79% (0.56–1.08)	1.19% (0.71–1.87)	0.47% (0.23–0.85)	1.20% (0.44–2.61)	0.70% (0.07–2.71)
Did someone from work repeatedly tell you about their sexual activities in a way that made you uncomfortable, angry, or upset? (Follow-up questions established that actions were also persistent or severe.)	1.37% (1.02–1.81)	2.22% (1.41–3.33)	1.03% (0.63–1.57)	0.82% (0.30–1.78)	0.92% (0.17–2.78)
Did someone from work repeatedly ask you questions about your sex life or sexual interests that made you uncomfortable, angry, or upset? (Follow-up questions established that actions were also persistent or severe.)	1.16% (0.86–1.52)	2.01% (1.37–2.85)	0.74% (0.40–1.26)	1.40% (0.46–3.20)	0.03% (0.00–0.82)
Did someone from work make repeated sexual comments about your appearance or body that made you uncomfortable, angry, or upset? (Follow-up questions established that actions were also persistent or severe.)	1.24% (0.92–1.64)	1.94% (1.23–2.89)	0.92% (0.52–1.49)	1.25% (0.57–2.35)	0.46% (0.07–1.55)
Did someone from work either take or share sexually suggestive pictures or videos of you when you did not want them to? (Follow-up questions established that actions were also persistent or severe.)	0.16% (0.06–0.34)	0.21% (0.05–0.59)	0.18% (0.04–0.53)	0.00% (0.00–0.96)	0.00% (0.00–1.12)

Table B.6.c—Continued

	Total	E1–E4	E5–E9	O1–O3	O4–O6
Did someone from work make repeated attempts to establish an unwanted romantic or sexual relationship with you? (Follow-up questions established that actions were also persistent or severe.)	0.72% (0.51–0.99)	0.84% (0.45–1.44)	0.63% (0.33–1.10)	1.03% (0.43–2.06)	0.37% (0.04–1.41)
Did someone from work intentionally touch you in a sexual way when you did not want them to? (Categorized as severe without requiring additional follow-up questions)	0.39% (0.24–0.59)	0.55% (0.25–1.07)	0.37% (0.15–0.76)	0.22% (0.02–0.90)	0.00% (0.00–1.12)
Did someone from work repeatedly touch you in any other way that made you uncomfortable, angry, or upset? (Follow-up questions established that actions were also persistent or severe.)	1.47% (1.09–1.94)	2.38% (1.56–3.47)	1.12% (0.65–1.78)	1.40% (0.57–2.87)	0.06% (0.00–0.88)
Has someone from work made you feel as if you would get some workplace benefit in exchange for doing something sexual? (Follow-up questions established that target had direct evidence of an exchange.)	0.05% (0.01–0.16)	0.05% (0.00–0.34)	0.06% (0.00–0.25)	0.00% (0.00–0.96)	0.13% (0.00–1.00)
Has someone from work made you feel like you would get punished or treated unfairly in the workplace if you did not do something sexual? (Follow-up questions established that target had direct evidence of an exchange.)	0.04% (0.00–0.14)	0.07% (0.00–0.37)	0.03% (0.00–0.20)	0.00% (0.00–0.96)	0.00% (0.00–1.12)
Did you hear someone from work say that [men/women] are not as good as [women/men] at your particular job, or that [men/women] should be prevented from having your job? (Follow-up question established that action resulted in harm to the target's career)	1.33% (0.95–1.81)	1.54% (0.97–2.31)	1.05% (0.49–1.96)	2.24% (1.29–3.60)	1.14% (0.40–2.55)
Do you think someone from work mistreated, ignored, excluded, or insulted you because you are a [man/woman]? (Follow-up question established that action resulted in harm to the target's career)	2.47% (1.98–3.04)	2.58% (1.84–3.52)	2.01% (1.31–2.96)	4.51% (2.89–6.68)	2.30% (1.16–4.06)

NOTE: Includes estimates for active-component Coast Guard members. 95-percent confidence intervals for each estimate are included in parentheses.

Too few warrant officers were included in the sample to break them out as a separate pay grade. For the purposes of this table, warrant officers have been included with the E5–E9 category.

B.7. Percentage who labeled the event(s) as sexual harassment among members who experienced sexual harassment in the past year

Table B.7.a
Percentage who labeled the event(s) as sexual harassment among members who experienced sexual harassment in the past year, by gender

SHFU1: How many of these behaviors that you selected as happening to you, do you consider to have been sexual harassment?

	Total	Men	Women
None were sexual harassment	50.78% (43.79–57.74)	67.20% (55.48–77.53)	32.12% (26.42–38.24)
Some were sexual harassment; some were not sexual harassment	42.27% (35.60–49.15)	29.58% (19.50–41.35)	56.69% (50.27–62.95)
All were sexual harassment	6.95% (4.81–9.67)	3.22% (0.76–8.65)	11.19% (7.71–15.54)

NOTE: Includes estimates for active-component Coast Guard members. 95-percent confidence intervals for each estimate are included in parentheses.

Table B.7.b
Percentage who labeled the event(s) as sexual harassment among members who experienced sexual harassment in the past year, by service

SHFU1: How many of these behaviors that you selected as happening to you, do you consider to have been sexual harassment?

	Total DoD	Army	Navy	Air Force	Marine Corps	Coast Guard
None were sexual harassment	42.41% (39.69–45.16)	40.74% (36.59–44.98)	40.63% (35.44–45.98)	44.83% (40.72–48.98)	50.29% (40.94–59.62)	50.78% (43.79–57.74)
Some were sexual harassment; some were not sexual harassment	47.84% (45.09–50.59)	48.40% (44.25–52.56)	50.17% (44.79–55.55)	44.21% (39.78–48.70)	43.67% (34.42–53.25)	42.27% (35.60–49.15)
All were sexual harassment	9.76% (8.23–11.46)	10.87% (8.75–13.29)	9.20% (5.78–13.72)	10.96% (9.12–13.03)	6.04% (3.87–8.94)	6.95% (4.81–9.67)

NOTE: Includes estimates for active-component DoD and Coast Guard members. 95-percent confidence intervals for each estimate are included in parentheses.

Table B.7.c
Percentage who labeled the event(s) as sexual harassment among members who experienced sexual harassment in the past year, by pay grade

SHFU1: How many of these behaviors that you selected as happening to you, do you consider to have been sexual harassment?

	Total	E1–E4	E5–E9	O1–O3	O4–O6
None were sexual harassment	50.78% (43.79–57.74)	43.47% (33.09–54.29)	55.32% (43.70–66.52)	NR (44.70–74.26)	NR (35.09–84.16)
Some were sexual harassment; some were not sexual harassment	42.27% (35.60–49.15)	50.14% (39.58–60.69)	36.28% (26.06–47.50)	33.55% (20.79–48.35)	NR (15.84–64.91)
All were sexual harassment	6.95% (4.81–9.67)	6.39% (3.35–10.89)	8.41% (4.66–13.73)	6.29% (2.09–14.04)	NR (0.00–27.14)

NOTE: Includes estimates for active-component Coast Guard members. 95-percent confidence intervals for each estimate are included in parentheses.

Too few warrant officers were included in the sample to break them out as a separate pay grade. For the purposes of this table, warrant officers have been included with the E5–E9 category.

NR = Not reportable.

B.8. Number of offenders involved in the sexual harassment or gender discrimination among members who experienced sexual harassment or gender discrimination in the past year

Table B.8.a
Number of offenders involved in the sexual harassment or gender discrimination among members who experienced sexual harassment or gender discrimination in the past year, by gender

Derived variable combining responses by those who had one experience (SHFU2) with responses by those who had more than one experience (SHFU2a). "Group" corresponds to SHFU2 response 'b', or SHFU2_1 response 'b'. Individual corresponds to SHFU2 response 'a' or SHFU2_1 response 'a'.

SHFU2: [Was this upsetting experience/were these upsetting experiences] the result of behavior by:

- One person
- A group of people in the same setting
- Different people in different situations [if 'c', show SHFU2_1]

SHFU2_1: You indicated that you had several situations like this happen to you. For the next series of questions that refer to the "upsetting situation" please think about the situation since [X date] that had the biggest effect on you–the one you consider to be the worst or the most serious.

Was the upsetting behavior in that situation due to:

a. One person
b. A group of people

	Total	Men	Women
Group	32.70%	34.56%	30.60%
	(26.58–39.28)	(24.17–46.17)	(25.48–36.09)
Individual	67.30%	65.44%	69.40%
	(60.72–73.42)	(53.83–75.83)	(63.91–74.52)

NOTE: Includes estimates for active-component Coast Guard members. 95-percent confidence intervals for each estimate are included in parentheses.

For respondents who experienced sexual harassment or gender discrimination behaviors involving multiple offenders across multiple situations, this question refers to their "worst or most serious" situation.

Table B.8.b
Number of offenders involved in the sexual harassment or gender discrimination among members who experienced sexual harassment or gender discrimination in the past year, by service

Derived variable combining responses by those who had one experience (SHFU2) with responses by those who had more than one experience (SHFU2a). "Group" corresponds to SHFU2 response 'b', or SHFU2_1 response 'b'. Individual corresponds to SHFU2 response 'a' or SHFU2_1 response 'a'.

SHFU2: [Was this upsetting experience/were these upsetting experiences] the result of behavior by:

- One person
- A group of people in the same setting
- Different people in different situations [if 'c', show SHFU2_1]

SHFU2_1: You indicated that you had several situations like this happen to you. For the next series of questions that refer to the "upsetting situation" please think about the situation since [X date] that had the biggest effect on you–the one you consider to be the worst or the most serious.

Was the upsetting behavior in that situation due to:

a.   One person
b.   A group of people

	Total DoD	Army	Navy	Air Force	Marine Corps	Coast Guard
Group	42.62%	38.12%	45.53%	39.80%	54.71%	32.70%
	(40.19–45.09)	(34.63–41.71)	(40.73–50.40)	(36.05–43.65)	(46.31–62.91)	(26.58–39.28)
Individual	57.38%	61.88%	54.47%	60.20%	45.29%	67.30%
	(54.91–59.81)	(58.29–65.37)	(49.60–59.27)	(56.35–63.95)	(37.09–53.69)	(60.72–73.42)

NOTE: Includes estimates for active-component DoD and Coast Guard members. 95-percent confidence intervals for each estimate are included in parentheses.

For respondents who experienced sexual harassment or gender discrimination behaviors involving multiple offenders across multiple situations, this question refers to their "worst or most serious" situation.

Table B.8.c
Number of offenders involved in the sexual harassment or gender discrimination among members who experienced sexual harassment or gender discrimination in the past year, by pay grade

Derived variable combining responses by those who had one experience (SHFU2) with responses by those who had more than one experience (SHFU2a). "Group" corresponds to SHFU2 response 'b', or SHFU2_1 response 'b'. Individual corresponds to SHFU2 response 'a' or SHFU2_1 response 'a'.

SHFU2: [Was this upsetting experience/were these upsetting experiences] the result of behavior by:

- One person
- A group of people in the same setting
- Different people in different situations [if 'c', show SHFU2_1]

SHFU2_1: You indicated that you had several situations like this happen to you. For the next series of questions that refer to the "upsetting situation" please think about the situation since [X date] that had the biggest effect on you–the one you consider to be the worst or the most serious.

Was the upsetting behavior in that situation due to:

a.   One person
b.   A group of people

	Total	E1–E4	E5–E9	O1–O3	O4–O6
Group	32.70% (26.58–39.28)	27.41% (19.59–36.41)	41.16% (29.97–53.06)	19.01% (9.65–31.93)	NR (24.36–65.75)
Individual	67.30% (60.72–73.42)	72.59% (63.59–80.41)	58.84% (46.94–70.03)	80.99% (68.07–90.35)	NR (34.25–75.64)

NOTE: Includes estimates for active-component Coast Guard members. 95-percent confidence intervals for each estimate are included in parentheses.

Too few warrant officers were included in the sample to break them out as a separate pay grade. For the purposes of this table, warrant officers have been included with the E5–E9 category.

For respondents who experienced sexual harassment or gender discrimination behaviors involving multiple offenders across multiple situations, this question refers to their "worst or most serious" situation.

NR = Not reportable.

B.9. Gender of the offender(s) among members who experienced sexual harassment or gender discrimination in the past year

Table B.9.a
Gender of the offender(s) among members who experienced sexual harassment or gender discrimination in the past year, by gender of the target

Derived variable combining responses of targets for whom the source of sexual harassment or gender discrimination was an individual (SHFU3a) with responses of targets for whom the source was a group (SHFU3b).

SHFU3a: Was this person a...

    a.   Man
    b.   Woman

SHFU3b: Were these people...

    a.   Men
    b.   Women
    c.   A mix of men and women

	Total	Men	Women
Man or men only	80.52% (73.56–86.32)	69.68% (57.83–79.88)	92.69% (89.59–95.11)
Woman or women only	10.83% (6.67–16.35)	18.25% (10.68–28.16)	2.48% (1.16–4.61)
A mix of men and women	8.66% (4.44–14.89)	12.07% (4.80–23.79)	4.83% (2.91–7.47)

NOTE: Includes estimates for active-component Coast Guard members. 95-percent confidence intervals for each estimate are included in parentheses.

For respondents who experienced sexual harassment or gender discrimination behaviors involving multiple offenders across multiple situations, this question refers to their "worst or most serious" situation.

Table B.9.b
Gender of the offender(s) among members who experienced sexual harassment or gender discrimination in the past year, by service

Derived variable combining responses of targets for whom the source of sexual harassment or gender discrimination was an individual (SHFU3a) with responses of targets for whom the source was a group (SHFU3b).

SHFU3a: Was this person a...

    a.   Man
    b.   Woman

SHFU3b: Were these people...

    a.   Men
    b.   Women
    c.   A mix of men and women

	Total DoD	Army	Navy	Air Force	Marine Corps	Coast Guard
Man or men only	75.00% (72.79–77.12)	75.45% (72.28–78.44)	72.04% (67.22–76.50)	74.60% (71.09–77.90)	81.97% (74.56–87.98)	80.52% (73.56–86.32)
Woman or women only	11.31% (9.72–13.05)	12.15% (9.92–14.69)	11.78% (8.30–16.06)	13.04% (10.16–16.38)	4.77% (2.75–7.64)	10.83% (6.67–16.35)
A mix of men and women	13.69% (12.08–15.44)	12.39% (10.23–14.83)	16.19% (12.85–19.99)	12.36% (10.35–14.60)	13.25% (7.69–20.77)	8.66% (4.44–14.89)

NOTE: Includes estimates for active-component DoD and Coast Guard members. 95-percent confidence intervals for each estimate are included in parentheses.

For respondents who experienced sexual harassment or gender discrimination behaviors involving multiple offenders across multiple situations, this question refers to their "worst or most serious" situation.

Table B.9.c
Gender of the offender(s) among members who experienced sexual harassment or gender discrimination in the past year, by pay grade

Derived variable combining responses of targets for whom the source of sexual harassment or gender discrimination was an individual (SHFU3a) with responses of targets for whom the source was a group (SHFU3b).

SHFU3a: Was this person a...

    a.   Man
    b.   Woman

SHFU3b: Were these people...

    a.   Men
    b.   Women
    c.   A mix of men and women

	Total	E1–E4	E5–E9	O1–O3	O4–O6
Man or men only	80.52% (73.56–86.32)	87.18% (78.34–93.36)	70.99% (57.80–82.01)	NR (70.40–96.35)	NR (71.06–96.44)
Woman or women only	10.83% (6.67–16.35)	7.32% (2.48–16.04)	15.75% (8.12–26.47)	NR (0.87–22.09)	NR (2.12–25.48)
A mix of men and women	8.66% (4.44–14.89)	5.50% (2.20–11.12)	13.26% (4.80–27.26)	6.07% (0.66–21.02)	2.83% (0.09–14.28)

NOTE: Includes estimates for active-component Coast Guard members. 95-percent confidence intervals for each estimate are included in parentheses.

Too few warrant officers were included in the sample to break them out as a separate pay grade. For the purposes of this table, warrant officers have been included with the E5–E9 category.

For respondents who experienced sexual harassment or gender discrimination behaviors involving multiple offenders across multiple situations, this question refers to their "worst or most serious" situation.

NR = Not reportable.

B.10. Workplace role of the offender(s) among members who experienced sexual harassment or gender discrimination in the past year

Table B.10.a
Workplace role of the offender(s) among members who experienced sexual harassment or gender discrimination in the past year, by gender of the target

Derived variable combining responses of targets for whom the offender was an individual (SHFU4) with responses of targets who were sexually harassed or discriminated against by a group (SHFU4a–SHFU4c).

For groups, offender role is coded according to the highest role played by any member of the group.

SHFU4: Was this person...

- One of your work supervisors or one of your unit leaders?
- One of your peers at about the same level?
- One of your subordinates or someone you managed?

Were any of the individuals who acted this way:

SHFU4a: One of your work supervisors or one of your unit leaders?		Yes	No
SHFU4b: One of your peers at about the same level?		Yes	No
SHFU4c: One of your subordinates or someone you managed?		Yes	No

	Total	Men	Women
Supervisor or leader	56.31%	52.74%	60.33%
	(49.79–62.67)	(41.52–63.77)	(54.61–65.85)
Peer	34.55%	36.12%	32.78%
	(28.54–40.95)	(25.94–47.31)	(27.51–38.39)
Subordinate	6.70%	7.39%	5.92%
	(4.06–10.30)	(3.18–14.21)	(3.66–8.98)
Other	2.44%	3.74%	0.97%
	(0.90–5.24)	(1.07–9.12)	(0.25–2.53)

NOTE: Includes estimates for active-component Coast Guard members. 95-percent confidence intervals for each estimate are included in parentheses.

For respondents who experienced sexual harassment or gender discrimination behaviors involving multiple offenders across multiple situations, this question refers to their "worst or most serious" situation.

Table B.10.b
Workplace role of the offender(s) among members who experienced sexual harassment or gender discrimination in the past year, by service

Derived variable combining responses of targets for whom the offender was an individual (SHFU4) with responses of targets who were sexually harassed or discriminated against by a group (SHFU4a–SHFU4c).

For groups, offender role is coded according to the highest role played by any member of the group.

SHFU4: Was this person...

- One of your work supervisors or one of your unit leaders?
- One of your peers at about the same level?
- One of your subordinates or someone you managed?

Were any of the individuals who acted this way:

SHFU4a: One of your work supervisors or one of your unit leaders?   Yes   No

SHFU4b: One of your peers at about the same level?   Yes   No

SHFU4c: One of your subordinates or someone you managed?   Yes   No

	Total DoD	Army	Navy	Air Force	Marine Corps	Coast Guard
Supervisor or leader	59.36% (57.00–61.68)	61.86% (58.14–65.48)	54.97% (50.27–59.61)	53.81% (50.20–57.40)	68.94% (61.99–75.31)	56.31% (49.79–62.67)
Peer	34.97% (32.71–37.28)	33.07% (29.46–36.84)	37.80% (33.40–42.36)	40.33% (36.93–43.81)	27.63% (21.65–34.27)	34.55% (28.54–40.95)
Subordinate	4.56% (3.96–5.23)	4.04% (3.32–4.87)	6.02% (4.56–7.77)	5.05% (3.99–6.30)	1.87% (0.96–3.27)	6.70% (4.06–10.30)
Other	1.11% (0.76–1.57)	1.02% (0.60–1.63)	1.21% (0.51–2.40)	0.80% (0.39–1.44)	1.56% (0.47–3.78)	2.44% (0.90–5.24)

NOTE: Includes estimates for active-component DoD and Coast Guard members. 95-percent confidence intervals for each estimate are included in parentheses.

For respondents who experienced sexual harassment or gender discrimination behaviors involving multiple offenders across multiple situations, this question refers to their "worst or most serious" situation.

Table B.10.c
Workplace role of the offender(s) among members who experienced sexual harassment or gender discrimination in the past year, by pay grade

Derived variable combining responses of targets for whom the offender was an individual (SHFU4) with responses of targets who were sexually harassed or discriminated against by a group (SHFU4a–SHFU4c).

For groups, offender role is coded according to the highest role played by any member of the group.

SHFU4: Was this person...

- One of your work supervisors or one of your unit leaders?
- One of your peers at about the same level?
- One of your subordinates or someone you managed?

Were any of the individuals who acted this way:

SHFU4a: One of your work supervisors or one of your unit leaders?   Yes   No
SHFU4b: One of your peers at about the same level?   Yes   No
SHFU4c: One of your subordinates or someone you managed?   Yes   No

	Total	E1–E4	E5–E9	O1–O3	O4–O6
Supervisor or leader	56.31% (49.79–62.67)	64.07% (54.15–73.17)	52.28% (40.97–63.41)	46.72% (32.29–61.55)	NR (29.99–71.35)
Peer	34.55% (28.54–40.95)	34.68% (25.64–44.60)	35.91% (25.65–47.22)	32.98% (20.39–47.66)	NR (9.00–49.36)
Subordinate	6.70% (4.06–10.30)	0.71% (0.07–2.77)	8.58% (4.11–15.38)	NR (8.58–36.07)	5.31% (0.87–16.22)
Other	2.44% (0.90–5.24)	0.55% (0.01–3.00)	3.24% (0.61–9.53)	0.52% (0.00–4.58)	NR (3.51–47.03)

NOTE: Includes estimates for active-component Coast Guard members. 95-percent confidence intervals for each estimate are included in parentheses.

Too few warrant officers were included in the sample to break them out as a separate pay grade. For the purposes of this table, warrant officers have been included with the E5–E9 category.

For respondents who experienced sexual harassment or gender discrimination behaviors involving multiple offenders across multiple situations, this question refers to their "worst or most serious" situation.

NR = Not reportable.

B.11. Military status of the offender(s) among members who experienced sexual harassment or gender discrimination in the past year

Table B.11.a
Military status of the offender(s) among members who experienced sexual harassment or gender discrimination in the past year, by gender of the target

Derived variable combining responses of targets for whom the offender was an individual (SHFU5) with responses of targets who were sexually harassed or discriminated against by a group (SHFU5a–SHFU5d).

Group sources are coded as military if at least one member of the group was in the military. They are coded as civilian/contractor if no group members are military and at least one member is a civilian or contractor. Groups are coded as 'neither' if all options received a 'no' response.

SHFU5: At the time of the upsetting situation was this person...

    a.   A Service member of higher rank than you?
    b.   A Service member of about the same rank as you?
    c.   A Service member of lower rank than you?
    d.   A civilian/contractor working for the military?
    e.   Do not know

At the time of the upsetting situation, were any of these individuals...

SHFU5a: Service members of higher rank than you?	Yes	No	Do not know
SHFU5b: Service members of about the same rank as you?	Yes	No	Do not know
SHFU5c: Service members of lower rank than you?	Yes	No	Do not know
SHFU5d: Civilians/contractors working for the military?	Yes	No	Do not know

	Total	Men	Women
Uniformed military	90.07% (85.07–93.83)	88.62% (79.07–94.84)	91.71% (88.24–94.41)
DoD civilian employee or contractor	8.51% (4.98–13.39)	9.85% (4.05–19.27)	7.01% (4.50–10.31)
Neither	0.00% (0.00–0.00)	0.00% (0.00–0.00)	0.00% (0.00–0.00)
Don't know	1.42% (0.38–3.65)	1.53% (0.11–6.42)	1.29% (0.43–2.93)

NOTE: Includes estimates for active-component Coast Guard members. 95-percent confidence intervals for each estimate are included in parentheses.

For respondents who experienced sexual harassment or gender discrimination behaviors involving multiple offenders across multiple situations, this question refers to their "worst or most serious" situation.

Table B.11.b
Military status of the offender(s) among members who experienced sexual harassment or gender discrimination in the past year, by service

Derived variable combining responses of targets for whom the offender was an individual (SHFU5) with responses of targets who were sexually harassed or discriminated against by a group (SHFU5a–SHFU5d).

Group sources are coded as military if at least one member of the group was in the military. They are coded as civilian/contractor if no group members are military and at least one member is a civilian or contractor. Groups are coded as 'neither' if all options 'no' response.

SHFU5: At the time of the upsetting situation was this person...

a.   A Service member of higher rank than you?
b.   A Service member of about the same rank as you?
c.   A Service member of lower rank than you?
d.   A civilian/contractor working for the military?
e.   Do not know

At the time of the upsetting situation, were any of these individuals...

SHFU5a: Service members of higher rank than you?	Yes	No	Do not know
SHFU5b: Service members of about the same rank as you?	Yes	No	Do not know
SHFU5c: Service members of lower rank than you?	Yes	No	Do not know
SHFU5d: Civilians/contractors working for the military?	Yes	No	Do not know

	Total DoD	Army	Navy	Air Force	Marine Corps	Coast Guard
Uniformed military	94.26% (93.32–95.11)	94.44% (92.77–95.82)	95.20% (93.39–96.63)	90.25% (88.47–91.84)	95.88% (92.83–97.89)	90.07% (85.07–93.83)
DoD civilian employee or contractor	3.12% (2.65–3.65)	2.94% (2.10–4.01)	2.04% (1.48–2.74)	7.60% (6.22–9.18)	1.32% (0.60–2.51)	8.51% (4.98–13.39)
Neither	0.13% (0.04–0.32)	0.10% (0.02–0.27)	0.20% (0.01–0.96)	0.13% (0.01–0.59)	0.05% (0.00–0.64)	0.00% (0.00–0.00)
Don't know	2.49% (1.82–3.32)	2.52% (1.49–3.97)	2.56% (1.35–4.38)	2.01% (1.32–2.93)	2.76% (1.01–5.94)	1.42% (0.38–3.65)

NOTE: Includes estimates for active-component DoD and Coast Guard members. 95-percent confidence intervals for each estimate are included in parentheses.

For respondents who experienced sexual harassment or gender discrimination behaviors involving multiple offenders across multiple situations, this question refers to their "worst or most serious" situation.

Table B.11.c
Military status of the offender(s) among members who experienced sexual harassment or gender discrimination in the past year, by pay grade

Derived variable combining responses of targets for whom the offender was an individual (SHFU5) with responses of targets who were sexually harassed or discriminated against by a group (SHFU5a–SHFU5d).

Group sources are coded as military if at least one member of the group was in the military. They are coded as civilian/contractor if no group members are military and at least one member is a civilian or contractor. Groups are coded as 'neither' if all options 'no' response.

SHFU5: At the time of the upsetting situation was this person...

a.   A Service member of higher rank than you?
b.   A Service member of about the same rank as you?
c.   A Service member of lower rank than you?
d.   A civilian/contractor working for the military?
e.   Do not know

At the time of the upsetting situation, were any of these individuals...

SHFU5a: Service members of higher rank than you?   Yes   No   Do not know

SHFU5b: Service members of about the same rank as you?   Yes   No   Do not know

SHFU5c: Service members of lower rank than you?   Yes   No   Do not know

SHFU5d: Civilians/contractors working for the military?   Yes   No   Do not know

	Total	E1–E4	E5–E9	O1–O3	O4–O6
Uniformed military	90.07% (85.07–93.83)	95.12% (87.31–98.80)	87.55% (77.16–94.37)	87.72% (78.12–94.16)	NR (49.75–90.24)
DoD civilian employee or contractor	8.51% (4.98–13.39)	4.12% (0.72–12.38)	10.22% (4.02–20.44)	11.12% (5.04–20.45)	NR (9.20–49.79)
Neither	0.00% (0.00–0.00)	0.00% (0.00–0.00)	0.00% (0.00–0.00)	0.00% (0.00–0.00)	0.00% (0.00–0.00)
Don't know	1.42% (0.38–3.65)	0.77% (0.08–2.84)	2.23% (0.24–8.12)	1.16% (0.03–6.22)	0.69% (0.00–9.17)

NOTE: Includes estimates for active-component Coast Guard members. 95-percent confidence intervals for each estimate are included in parentheses.

Too few warrant officers were included in the sample to break them out as a separate pay grade. For the purposes of this table, warrant officers have been included with the E5–E9 category.

For respondents who experienced sexual harassment or gender discrimination behaviors involving multiple offenders across multiple situations, this question refers to their "worst or most serious" situation.

NR = Not reportable.

B.12. Military rank of the offender(s) among members who experienced sexual harassment or gender discrimination in the past year by another service member

Table B.12.a
Military rank of the offender(s) among members who experienced sexual harassment or gender discrimination in the past year by another service member, by gender of the target

Derived variable combining responses of targets for whom the offender was an individual (SHFU5) with responses of targets who were sexually harassed or discriminated against by a group (SHFU5a–SHFU5d).

Coded only for targets with at least one military offender. For groups, rank is coded according to the highest ranking member of the group.

SHFU5: At the time of the upsetting situation was this person...

  a. A Service member of higher rank than you?
  b. A Service member of about the same rank as you?
  c. A Service member of lower rank than you?
  d. A civilian/contractor working for the military?
  e. Do not know

At the time of the upsetting situation, were any of these individuals...

SHFU5a: Service members of higher rank than you?   Yes   No   Do not know

SHFU5b: Service members of about the same rank as you?   Yes   No   Do not know

SHFU5c: Service members of lower rank than you?   Yes   No   Do not know

SHFU5d: Civilians/contractors working for the military?   Yes   No   Do not know

	Total	Men	Women
Higher rank	65.82%	61.48%	70.53%
	(59.21–72.00)	(49.84–72.23)	(64.69–75.91)
Similar rank	24.39%	26.33%	22.28%
	(18.98–30.48)	(17.23–37.19)	(17.34–27.88)
Lower rank	9.80%	12.19%	7.19%
	(6.19–14.54)	(6.04–21.20)	(4.60–10.62)

NOTE: Includes estimates for active-component Coast Guard members. 95-percent confidence intervals for each estimate are included in parentheses.

For respondents who experienced sexual harassment or gender discrimination behaviors involving multiple offenders across multiple situations, this question refers to their "worst or most serious" situation.

Table B.12.b
Military rank of the offender(s) among members who experienced sexual harassment or gender discrimination in the past year by another service member, by service

Derived variable combining responses of targets for whom the offender was an individual (SHFU5) with responses of targets who were sexually harassed or discriminated against by a group (SHFU5a–SHFU5d).

Coded only for targets with at least one military offender. For groups, rank is coded according to the highest ranking member of the group.

SHFU5: At the time of the upsetting situation was this person...

    a.   A Service member of higher rank than you?
    b.   A Service member of about the same rank as you?
    c.   A Service member of lower rank than you?
    d.   A civilian/contractor working for the military?
    e.   Do not know

At the time of the upsetting situation, were any of these individuals...

SHFU5a: Service members of higher rank than you?	Yes	No Do not know
SHFU5b: Service members of about the same rank as you?	Yes	No Do not know
SHFU5c: Service members of lower rank than you?	Yes	No Do not know
SHFU5d: Civilians/contractors working for the military?	Yes	No Do not know

	Total DoD	Army	Navy	Air Force	Marine Corps	Coast Guard
Higher rank	66.98% (64.60–69.30)	68.02% (64.09–71.76)	63.37% (58.70–67.86)	64.72% (61.23–68.09)	75.75% (69.60–81.20)	65.82% (59.21–72.00)
Similar rank	26.74% (24.52–29.04)	26.28% (22.59–30.23)	28.60% (24.44–33.06)	28.89% (25.82–32.12)	20.78% (15.81–26.49)	24.39% (18.98–30.48)
Lower rank	6.28% (5.38–7.29)	5.71% (4.39–7.27)	8.03% (6.09–10.33)	6.39% (5.04–7.96)	3.48% (1.82–5.97)	9.80% (6.19–14.54)

NOTE: Includes estimates for active-component DoD and Coast Guard members. 95-percent confidence intervals for each estimate are included in parentheses.

For respondents who experienced sexual harassment or gender discrimination behaviors involving multiple offenders across multiple situations, this question refers to their "worst or most serious" situation.

Table B.12.c
Military rank of the offender(s) among members who experienced sexual harassment or gender discrimination in the past year by another service member, by pay grade

Derived variable combining responses of targets for whom the offender was an individual (SHFU5) with responses of targets who were sexually harassed or discriminated against by a group (SHFU5a–SHFU5d).

Coded only for targets with at least one military offender. For groups, rank is coded according to the highest ranking member of the group.

SHFU5: At the time of the upsetting situation was this person...

    a.   A Service member of higher rank than you?
    b.   A Service member of about the same rank as you?
    c.   A Service member of lower rank than you?
    d.   A civilian/contractor working for the military?
    e.   Do not know

At the time of the upsetting situation, were any of these individuals...

SHFU5a: Service members of higher rank than you?	Yes	No	Do not know
SHFU5b: Service members of about the same rank as you?	Yes	No	Do not know
SHFU5c: Service members of lower rank than you?	Yes	No	Do not know
SHFU5d: Civilians/contractors working for the military?	Yes	No	Do not know

	Total	E1–E4	E5–E9	O1–O3	O4–O6
Higher rank	65.82% (59.21–72.00)	76.48% (67.14–84.26)	60.30% (48.50–71.29)	NR (32.70–64.89)	NR (34.64–82.56)
Similar rank	24.39% (18.98–30.48)	23.27% (15.50–32.63)	26.14% (16.97–37.13)	NR (11.99–39.87)	NR (7.20–39.91)
Lower rank	9.80% (6.19–14.54)	0.25% (0.00–2.05)	13.56% (6.83–23.24)	NR (14.05–44.48)	NR (2.30–55.43)

NOTE: Includes estimates for active-component Coast Guard members. 95-percent confidence intervals for each estimate are included in parentheses.

Too few warrant officers were included in the sample to break them out as a separate pay grade. For the purposes of this table, warrant officers have been included with the E5–E9 category.

For respondents who experienced sexual harassment or gender discrimination behaviors involving multiple offenders across multiple situations, this question refers to their "worst or most serious" situation.

NR = Not reportable.

B.13. Duration of the upsetting behavior among members who experienced sexual harassment or gender discrimination in the past year

Table B.13.a
Duration of the upsetting behavior among members who experienced sexual harassment or gender discrimination in the past year, by gender

SHFU6: Thinking about this situation, about how long did these upsetting behaviors continue? If the situation is still happening, indicate how long it has been going on.

	Total	Men	Women
It happened one time	22.77% (17.78–28.40)	24.62% (16.24–34.69)	20.68% (16.38–25.53)
About one week	7.95% (5.34–11.30)	8.17% (4.08–14.31)	7.71% (5.02–11.23)
About one month	9.17% (5.87–13.50)	9.80% (4.49–18.00)	8.46% (5.68–12.01)
A few months	32.79% (27.01–38.97)	26.88% (17.73–37.74)	39.44% (33.83–45.26)
A year or more	27.33% (21.30–34.03)	30.53% (20.24–42.47)	23.72% (19.22–28.69)

NOTE: Includes estimates for active-component Coast Guard members. 95-percent confidence intervals for each estimate are included in parentheses.

For respondents who experienced sexual harassment or gender discrimination behaviors involving multiple offenders across multiple situations, this question refers to their "worst or most serious" situation.

Table B.13.b
Duration of the upsetting behavior among members who experienced sexual harassment or gender discrimination in the past year, by service

SHFU6: Thinking about this situation, about how long did these upsetting behaviors continue? If the situation is still happening, indicate how long it has been going on.

	Total DoD	Army	Navy	Air Force	Marine Corps	Coast Guard
It happened one time	20.30% (18.55–22.15)	19.86% (17.22–22.70)	19.67% (16.61–23.01)	20.23% (17.69–22.96)	23.77% (16.84–31.91)	22.77% (17.78–28.40)
About one week	10.42% (9.08–11.89)	12.96% (10.30–16.01)	9.01% (7.29–10.99)	9.44% (7.76–11.35)	6.16% (4.20–8.66)	7.95% (5.34–11.30)
About one month	11.19% (9.96–12.52)	10.99% (9.04–13.20)	10.71% (8.57–13.18)	11.93% (10.08–13.99)	12.35% (8.31–17.43)	9.17% (5.87–13.50)
A few months	32.61% (30.43–34.85)	33.83% (30.51–37.27)	31.61% (27.51–35.93)	34.72% (30.78–38.82)	28.35% (21.24–36.34)	32.79% (27.01–38.97)
A year or more	25.47% (23.03–28.03)	22.36% (19.10–25.90)	29.00% (23.81–34.63)	23.67% (21.05–26.44)	29.38% (21.21–38.65)	27.33% (21.30–34.03)

NOTE: Includes estimates for active-component DoD and Coast Guard members. 95-percent confidence intervals for each estimate are included in parentheses.

For respondents who experienced sexual harassment or gender discrimination behaviors involving multiple offenders across multiple situations, this question refers to their "worst or most serious" situation.

Table B.13.c
Duration of the upsetting behavior among members who experienced sexual harassment or gender discrimination in the past year, by pay grade

SHFU6: Thinking about this situation, about how long did these upsetting behaviors continue? If the situation is still happening, indicate how long it has been going on.

	Total	E1–E4	E5–E9	O1–O3	O4–O6
It happened one time	22.77% (17.78–28.40)	23.49% (15.65–32.93)	19.16% (11.94–28.31)	27.15% (15.49–41.68)	NR (15.89–61.68)
About one week	7.95% (5.34–11.30)	9.15% (5.43–14.24)	7.13% (3.15–13.52)	8.12% (2.00–20.57)	3.95% (0.37–14.78)
About one month	9.17% (5.87–13.50)	9.01% (4.11–16.66)	8.87% (4.00–16.50)	11.39% (3.81–24.62)	6.63% (1.10–19.86)
A few months	32.79% (27.01–38.97)	39.81% (30.51–49.68)	30.44% (20.98–41.30)	20.96% (11.47–33.50)	NR (12.35–42.34)
A year or more	27.33% (21.30–34.03)	18.54% (10.92–28.47)	34.39% (23.56–46.56)	NR (19.30–47.84)	NR (13.85–45.41)

NOTE: Includes estimates for active-component Coast Guard members. 95-percent confidence intervals for each estimate are included in parentheses.

Too few warrant officers were included in the sample to break them out as a separate pay grade. For the purposes of this table, warrant officers have been included with the E5–E9 category.

For respondents who experienced sexual harassment or gender discrimination behaviors involving multiple offenders across multiple situations, this question refers to their "worst or most serious" situation.

NR = Not reportable.

B.14. Location of the upsetting behavior among members who experienced sexual harassment or gender discrimination in the past year

Table B.14.a
Location of the upsetting behavior among members who experienced sexual harassment or gender discrimination in the past year, by gender

SHFU7: Thinking about this upsetting behavior, did it ever occur...

	Total	Men	Women
On a military installation/ship?	90.93% (87.45–93.70)	91.66% (85.44–95.82)	90.11% (86.41–93.08)
While you were on TDY/TAD, at sea, or during field exercises/alerts?	25.24% (19.43–31.79)	26.86% (17.08–38.63)	23.42% (18.71–28.67)
While you were deployed to a combat zone or to an area where you drew imminent danger pay or hostile fire pay?	4.34% (1.03–11.52)	6.34% (0.87–20.29)	2.08% (0.76–4.50)
During recruit training/basic training?	6.12% (2.25–12.93)	8.60% (2.13–21.65)	3.31% (1.52–6.21)
In a civilian location?	27.82% (22.02–34.23)	24.21% (14.67–36.09)	31.90% (26.77–37.37)

NOTE: Includes estimates for active-component Coast Guard members. 95-percent confidence intervals for each estimate are included in parentheses.

For respondents who experienced sexual harassment or gender discrimination behaviors involving multiple offenders across multiple situations, this question refers to their "worst or most serious" situation.

Table B.14.b
Location of the upsetting behavior among members who experienced sexual harassment or gender discrimination in the past year, by service

SHFU7: Thinking about this upsetting behavior, did it ever occur...

	Total DoD	Army	Navy	Air Force	Marine Corps	Coast Guard
On a military installation/ship?	93.70% (92.59–94.68)	93.91% (92.49–95.13)	94.97% (93.41–96.26)	94.31% (92.85–95.56)	88.69% (81.11–94.01)	90.93% (87.45–93.70)
While you were on TDY/TAD, at sea, or during field exercises/ alerts?	31.34% (29.05–33.70)	31.85% (28.29–35.58)	34.95% (30.59–39.51)	19.36% (16.96–21.93)	34.06% (26.10–42.74)	25.24% (19.43–31.79)
While you were deployed to a combat zone or to an area where you drew imminent danger pay or hostile fire pay?	21.23% (19.29–23.28)	23.75% (20.71–27.01)	20.92% (17.35–24.85)	16.91% (14.69–19.32)	18.08% (11.46–26.45)	4.34% (1.03–11.52)
During recruit training/basic training?	11.73% (9.92–13.75)	13.78% (11.50–16.33)	8.88% (5.54–13.32)	6.03% (4.67–7.65)	18.90% (11.03–29.16)	6.12% (2.25–12.93)
In a civilian location?	23.54% (21.76–25.39)	24.56% (21.59–27.72)	20.79% (18.04–23.75)	24.80% (22.12–27.63)	25.82% (19.37–33.14)	27.82% (22.02–34.23)

NOTE: Includes estimates for active-component DoD and Coast Guard members. 95-percent confidence intervals for each estimate are included in parentheses.

For respondents who experienced sexual harassment or gender discrimination behaviors involving multiple offenders across multiple situations, this question refers to their "worst or most serious" situation.

Table B.14.c
Location of the upsetting behavior among members who experienced sexual harassment or gender discrimination in the past year, by pay grade

SHFU7: Thinking about this upsetting behavior, did it ever occur...

	Total	E1–E4	E5–E9	O1–O3	O4–O6
On a military installation/ship?	90.93% (87.45–93.70)	90.35% (85.24–94.14)	94.21% (88.78–97.51)	88.75% (76.45–95.98)	NR (43.10–91.81)
While you were on TDY/TAD, at sea, or during field exercises/alerts?	25.24% (19.43–31.79)	22.68% (14.98–32.02)	27.13% (16.75–39.73)	25.84% (14.44–40.27)	NR (12.13–53.01)
While you were deployed to a combat zone or to an area where you drew imminent danger pay or hostile fire pay?	4.34% (1.03–11.52)	0.89% (0.10–3.25)	NR (1.50–25.40)	2.70% (0.50–8.05)	0.00% (0.00–12.98)
During recruit training/basic training?	6.12% (2.25–12.93)	4.64% (1.34–11.20)	NR (1.77–25.59)	2.76% (0.53–8.13)	0.72% (0.00–9.07)
In a civilian location?	27.82% (22.02–34.23)	27.92% (19.48–37.69)	24.57% (14.84–36.66)	33.63% (21.00–48.23)	NR (20.66–62.00)

NOTE: Includes estimates for active-component Coast Guard members. 95-percent confidence intervals for each estimate are included in parentheses.

Too few warrant officers were included in the sample to break them out as a separate pay grade. For the purposes of this table, warrant officers have been included with the E5–E9 category.

For respondents who experienced sexual harassment or gender discrimination behaviors involving multiple offenders across multiple situations, this question refers to their "worst or most serious" situation.

NR = Not reportable.

B.15. Consequences of sexual harassment and discrimination among members who experienced sexual harassment or gender discrimination in the past year

Table B.15.a
Consequences of sexual harassment and discrimination among members who experienced sexual harassment or gender discrimination in the past year, by gender

SHFU8: Thinking about this upsetting situation...

	Total	Men	Women
Did you request a transfer or other change of assignment as a result of the situation?	11.12% (6.55–17.33)	9.88% (2.88–22.96)	12.52% (9.09–16.67)
Did it make you want to leave the military?	38.59% (32.30–45.17)	31.65% (21.25–43.59)	46.39% (40.68–52.17)
Did this situation ever make it hard to do your job or complete your work?	51.65% (45.11–58.15)	42.66% (31.70–54.17)	61.80% (56.16–67.21)
Did this situation ever make your workplace either less productive or compromise your unit's mission?	45.50% (39.02–52.10)	45.48% (34.39–56.91)	45.52% (39.85–51.28)
Did you take a sick call day or any other type of leave because of this situation?	14.19% (9.50–20.07)	9.56% (2.75–22.39)	19.38% (15.15–24.22)
Did this situation negatively affect your evaluation/fitness reports or promotions?	25.41% (19.86–31.62)	20.16% (11.29–31.83)	31.34% (26.19–36.86)
Did this situation either cause arguments in the workplace or damage unit cohesion?	50.08% (43.56–56.60)	45.99% (34.75–57.52)	54.63% (48.91–60.26)
Did this situation damage your relationship with coworkers?	52.88% (46.38–59.30)	49.02% (37.80–60.31)	57.20% (51.55–62.72)
Did this situation damage your other personal relationships, for example, with your spouse or a friend?	18.64% (13.73–24.41)	11.59% (4.40–23.45)	26.57% (21.70–31.90)

NOTE: Includes estimates for active-component Coast Guard members. 95-percent confidence intervals for each estimate are included in parentheses.

For respondents who experienced sexual harassment or gender discrimination behaviors involving multiple offenders across multiple situations, this question refers to their "worst or most serious" situation.

Table B.15.b

Consequences of sexual harassment and discrimination among members who experienced sexual harassment or gender discrimination in the past year, by service

SHFU8: Thinking about this upsetting situation...

	Total DoD	Army	Navy	Air Force	Marine Corps	Coast Guard
Did you request a transfer or other change of assignment as a result of the situation?	13.02% (11.50–14.67)	14.65% (12.45–17.07)	11.54% (8.39–15.36)	11.82% (10.07–13.74)	12.60% (8.15–18.32)	11.12% (6.55–17.33)
Did it make you want to leave the military?	42.46% (40.03–44.92)	43.10% (39.47–46.78)	41.54% (36.74–46.45)	35.75% (32.37–39.25)	50.79% (42.23–59.31)	38.59% (32.30–45.17)
Did this situation ever make it hard to do your job or complete your work?	50.27% (47.83–52.71)	51.11% (47.39–54.83)	50.79% (46.00–55.57)	49.97% (46.29–53.65)	46.13% (37.72–54.70)	51.65% (45.11–58.15)
Did this situation ever make your workplace either less productive or compromise your unit's mission?	48.09% (45.65–50.53)	52.26% (48.55–55.96)	46.78% (41.96–51.64)	44.43% (40.86–48.06)	40.84% (32.77–49.29)	45.50% (39.02–52.10)
Did you take a sick call day or any other type of leave because of this situation?	12.80% (11.30–14.43)	14.21% (12.05–16.61)	11.45% (8.22–15.40)	12.91% (11.16–14.83)	11.19% (7.46–15.94)	14.19% (9.50–20.07)
Did this situation negatively affect your evaluation/ fitness reports or promotions?	28.13% (25.91–30.45)	31.51% (28.29–34.87)	26.06% (21.49–31.06)	21.16% (18.16–24.42)	29.89% (22.34–38.35)	25.41% (19.86–31.62)
Did this situation either cause arguments in the workplace or damage unit cohesion?	52.73% (50.28–55.18)	55.04% (51.33–58.71)	52.31% (47.48–57.10)	50.05% (46.36–53.74)	48.70% (40.22–57.24)	50.08% (43.56–56.60)
Did this situation damage your relationship with coworkers?	53.76% (51.31–56.20)	53.07% (49.34–56.78)	55.22% (50.42–59.95)	51.88% (48.17–55.58)	54.60% (45.87–63.12)	52.88% (46.38–59.30)
Did this situation damage your other personal relationships, for example, with your spouse or a friend?	22.07% (20.17–24.06)	20.89% (18.58–23.34)	21.91% (17.91–26.33)	20.72% (18.38–23.21)	28.46% (20.94–36.99)	18.64% (13.73–24.41)

NOTE: Includes estimates for active-component DoD and Coast Guard members. 95-percent confidence intervals for each estimate are included in parentheses.

For respondents who experienced sexual harassment or gender discrimination behaviors involving multiple offenders across multiple situations, this question refers to their "worst or most serious" situation.

Table B.15.c
Consequences of sexual harassment and discrimination among members who experienced sexual harassment or gender discrimination in the past year, by pay grade

SHFU8: Thinking about this upsetting situation...

	Total	E1–E4	E5–E9	O1–O3	O4–O6
Did you request a transfer or other change of assignment as a result of the situation?	11.12% (6.55–17.33)	10.53% (5.57–17.66)	13.81% (5.11–28.03)	4.03% (1.02–10.41)	12.71% (4.24–27.26)
Did it make you want to leave the military?	38.59% (32.30–45.17)	44.13% (34.46–54.14)	34.94% (24.22–46.89)	36.80% (23.87–51.30)	NR (12.94–44.09)
Did this situation ever make it hard to do your job or complete your work?	51.65% (45.11–58.15)	50.01% (40.10–59.92)	51.07% (39.79–62.26)	NR (37.22–67.21)	NR (46.78–88.02)
Did this situation ever make your workplace either less productive or compromise your unit's mission?	45.50% (39.02–52.10)	42.79% (33.10–52.90)	47.85% (36.60–59.27)	41.65% (28.00–56.30)	NR (38.38–79.76)
Did you take a sick call day or any other type of leave because of this situation?	14.19% (9.50–20.07)	11.12% (6.01–18.34)	16.78% (7.85–29.74)	12.84% (6.54–21.92)	NR (10.74–39.12)
Did this situation negatively affect your evaluation/fitness reports or promotions?	25.41% (19.86–31.62)	21.17% (14.22–29.61)	28.53% (18.34–40.62)	26.74% (15.61–40.54)	NR (16.66–50.38)
Did this situation either cause arguments in the workplace or damage unit cohesion?	50.08% (43.56–56.60)	50.03% (40.12–59.93)	51.24% (39.92–62.47)	43.36% (29.39–58.16)	NR (38.13–79.44)
Did this situation damage your relationship with coworkers?	52.88% (46.38–59.30)	56.29% (46.45–65.78)	52.29% (40.98–63.43)	44.15% (30.19–58.83)	NR (32.64–72.61)
Did this situation damage your other personal relationships, for example, with your spouse or a friend?	18.64% (13.73–24.41)	14.42% (9.91–20.01)	22.60% (12.92–35.03)	20.69% (10.85–33.92)	14.63% (5.65–28.95)

NOTE: Includes estimates for active-component Coast Guard members. 95-percent confidence intervals for each estimate are included in parentheses.

Too few warrant officers were included in the sample to break them out as a separate pay grade. For the purposes of this table, warrant officers have been included with the E5–E9 category.

For respondents who experienced sexual harassment or gender discrimination behaviors involving multiple offenders across multiple situations, this question refers to their "worst or most serious" situation.

NR = Not reportable.

B.16. Disclosure among members who experienced sexual harassment or gender discrimination in the past year

Table B.16.a
Disclosure among members who experienced sexual harassment or gender discrimination in the past year, by gender

Variable derived from responses to SHFU9a–d. Three mutually exclusive categories were created. "Reported" includes all targets who discussed the situation either with a work supervisor or someone up the chain of command (SHFU9c) or with any person tasked with enforcing equal opportunity regulations (SHFU9d). "Disclosed to someone unofficially" includes targets who did not disclose to someone officially, but who did tell friends, family or coworkers (SHFU9a) or a chaplain, counselor, or medical person (SHFU9b). "Did not disclose to anyone" includes all targets who told no one about the event(s).

Thinking about this upsetting situation:

SHFU9a: Did you discuss this situation with your friends, family, or co-workers?   Yes   No

SHFU9b: Did you discuss this situation with a chaplain, counselor, or medical person?   Yes   No

SHFU9c: Did you discuss this situation with a work supervisor or anyone up your chain of command?   Yes   No

SHFU9d: Did you officially report this situation as possible harassment to any person tasked with enforcing sexual harassment or Equal Opportunity regulations?   Yes   No

	Total	Men	Women
Reported	42.60% (36.22–49.17)	37.89% (27.16–49.56)	47.86% (42.15–53.61)
Disclosed to someone unofficially	37.55% (31.58–43.81)	33.15% (23.41–44.07)	42.46% (36.90–48.18)
Did not disclose to anyone	19.85% (14.62–25.97)	28.97% (19.60–39.86)	9.67% (6.90–13.08)

NOTE: Includes estimates for active-component Coast Guard members.
95-percent confidence intervals for each estimate are included in parentheses.

For respondents who experienced sexual harassment or gender discrimination behaviors involving multiple offenders across multiple situations, this question refers to their "worst or most serious" situation.

Table B.16.b
Disclosure among members who experienced sexual harassment or gender discrimination in the past year, by service

Variable derived from responses to SHFU9a–d. Three mutually exclusive categories were created. "Reported" includes all targets who discussed the situation either with a work supervisor or someone up the chain of command (SHFU9c) or with any person tasked with enforcing equal opportunity regulations (SHFU9d). "Disclosed to someone unofficially" includes targets who did not disclose to someone officially, but who did tell friends, family or coworkers (SHFU9a) or a chaplain, counselor, or medical person (SHFU9b). "Did not disclose to anyone" includes all targets who told no one about the event(s).

Thinking about this upsetting situation:

SHFU9a: Did you discuss this situation with your friends, family, or co-workers?   Yes   No

SHFU9b: Did you discuss this situation with a chaplain, counselor, or medical person?   Yes   No

SHFU9c: Did you discuss this situation with a work supervisor or anyone up your chain of command?   Yes   No

SHFU9d: Did you officially report this situation as possible harassment to any person tasked with enforcing sexual harassment or Equal Opportunity regulations?   Yes   No

	Total DoD	Army	Navy	Air Force	Marine Corps	Coast Guard
Reported	37.92% (35.68–40.19)	39.05% (35.67–42.52)	37.75% (33.44–42.20)	37.73% (34.57–40.98)	34.44% (26.56–43.01)	42.60% (36.22–49.17)
Disclosed to someone unofficially	37.85% (35.43–40.31)	38.11% (34.41–41.91)	38.19% (33.46–43.09)	39.26% (35.36–43.28)	34.29% (26.85–42.35)	37.55% (31.58–43.81)
Did not disclose to anyone	24.23% (22.08–26.49)	22.84% (19.79–26.12)	24.07% (19.85–28.69)	23.00% (20.17–26.03)	31.27% (23.37–40.05)	19.85% (14.62–25.97)

NOTE: Includes estimates for active-component DoD and Coast Guard members. 95-percent confidence intervals for each estimate are included in parentheses.

For respondents who experienced sexual harassment or gender discrimination behaviors involving multiple offenders across multiple situations, this question refers to their "worst or most serious" situation.

Table B.16.c
Disclosure among members who experienced sexual harassment or gender discrimination in the past year, by pay grade

Variable derived from responses to SHFU9a–d. Three mutually exclusive categories were created. "Reported" includes all targets who discussed the situation either with a work supervisor or someone up the chain of command (SHFU9c) or with any person tasked with enforcing equal opportunity regulations (SHFU9d). "Disclosed to someone unofficially" includes targets who did not disclose to someone officially, but who did tell friends, family or coworkers (SHFU9a) or a chaplain, counselor, or medical person (SHFU9b). "Did not disclose to anyone" includes all targets who told no one about the event(s).

Thinking about this upsetting situation:

SHFU9a: Did you discuss this situation with your friends, family, or co-workers?   Yes   No

SHFU9b: Did you discuss this situation with a chaplain, counselor, or medical person?   Yes   No

SHFU9c: Did you discuss this situation with a work supervisor or anyone up your chain of command?   Yes   No

SHFU9d: Did you officially report this situation as possible harassment to any person tasked with enforcing sexual harassment or Equal Opportunity regulations?   Yes   No

	Total	E1–E4	E5–E9	O1–O3	O4–O6
Reported	42.60% (36.22–49.17)	38.86% (29.75–48.59)	48.42% (37.17–59.79)	34.49% (21.85–48.98)	NR (27.19–68.26)
Disclosed to someone unofficially	37.55% (31.58–43.81)	39.23% (29.90–49.18)	33.24% (23.75–43.84)	41.55% (27.90–56.21)	NR (29.65–70.73)
Did not disclose to anyone	19.85% (14.62–25.97)	21.90% (13.55–32.35)	18.34% (10.61–28.51)	NR (11.37–41.05)	2.38% (0.08–11.85)

NOTE: Includes estimates for active-component Coast Guard members. 95-percent confidence intervals for each estimate are included in parentheses.

Too few warrant officers were included in the sample to break them out as a separate pay grade. For the purposes of this table, warrant officers have been included with the E5–E9 category.

For respondents who experienced sexual harassment or gender discrimination behaviors involving multiple offenders across multiple situations, this question refers to their "worst or most serious" situation.

NR = Not reportable.

B.17. Leadership actions taken in response to the disclosure among members who experienced sexual harassment or gender discrimination in the past year and disclosed it to a supervisor, leader, or official

Table B.17.a

Leadership actions taken in response to the disclosure among members who experienced sexual harassment or gender discrimination in the past year and disclosed it to a supervisor, leader, or official, by gender

SHFU10: What actions were taken in response to your [discussing the situation with a supervisor or anyone up your chain of command/reporting the situation]?

	Total	Men	Women
No action was taken because you asked for the discussion to be kept private.	23.86% (14.30–35.82)	NR (11.65–51.99)	19.47% (13.13–27.21)
You discussed the situation, but no action was taken because you chose not to give enough details about the situation.	14.53% (6.03–27.71)	NR (5.90–46.67)	8.51% (4.43–14.49)
The person you told took no action.	42.31% (31.93–53.21)	NR (27.57–66.52)	38.44% (30.38–46.99)
The rules on harassment were explained to everyone in the workplace.	60.35% (50.37–69.74)	NR (46.03–80.86)	56.33% (47.58–64.80)
Someone talked to the [person|people] to ask them to change their behavior.	53.78% (43.58–63.76)	NR (35.78–73.04)	52.78% (44.15–61.30)
Your work station or duties were changed to help you avoid [that person|those people].	21.51% (12.32–33.40)	NR (4.52–46.27)	23.23% (16.44–31.21)
The [person was moved|people were moved] or reassigned so that you did not have as much contact with them.	19.90% (10.68–32.27)	NR (5.82–47.54)	18.44% (12.22–26.15)
There was some official career action taken against the [person|people] for their upsetting behavior (for example, a negative evaluation/fitness report).	16.04% (7.58–28.33)	NR (1.63–43.79)	17.59% (11.47–25.24)
The [person|people] stopped their upsetting behavior.	34.87% (24.65–46.22)	NR (19.31–59.46)	32.20% (24.40–40.82)
You were encouraged to drop the issue.	31.73% (21.53–43.40)	NR (14.49–56.22)	30.68% (23.35–38.79)
You were discouraged from filing a formal complaint.	29.46% (19.25–41.43)	NR (13.27–55.34)	27.58% (20.40–35.73)
The [person|people] who did this retaliated against you for complaining. For example, their upsetting behavior became worse or they threatened you.	28.40% (18.45–40.18)	NR (11.28–52.85)	27.99% (20.61–36.37)

Table B.17.a—Continued

	Total	Men	Women
Your coworkers treated you worse, avoided you, or blamed you for the problem.	24.02% (14.79–35.45)	NR (3.24–45.15)	29.77% (22.17–38.30)
Your supervisor punished you for bringing it up (e.g., loss of privileges, denied promotion/training, transferred to less favorable job).	20.62% (11.46–32.69)	NR (5.48–47.09)	20.28% (14.09–27.72)

NOTE: Includes estimates for active-component Coast Guard members. 95-percent confidence intervals for each estimate are included in parentheses.

For respondents who experienced sexual harassment or gender discrimination behaviors involving multiple offenders across multiple situations, this question refers to their "worst or most serious" situation.

NR = Not reportable.

Table B.17.b
Leadership actions taken in response to the disclosure among members who experienced sexual harassment or gender discrimination in the past year and disclosed it to a supervisor, leader, or official, by service

SHFU10: What actions were taken in response to your [discussing the situation with a supervisor or anyone up your chain of command/reporting the situation]?

	Total DoD	Army	Navy	Air Force	Marine Corps	Coast Guard
No action was taken because you asked for the discussion to be kept private.	27.55% (24.45–30.82)	26.74% (22.19–31.69)	28.13% (22.75–34.00)	24.10% (20.47–28.02)	NR (19.99–49.74)	23.86% (14.30–35.82)
You discussed the situation, but no action was taken because you chose not to give enough details about the situation.	14.57% (11.88–17.59)	15.92% (11.70–20.93)	11.27% (7.97–15.34)	11.86% (9.02–15.20)	NR (10.06–39.50)	14.53% (6.03–27.71)
The person you told took no action.	41.39% (37.95–44.89)	44.82% (39.66–50.06)	41.67% (34.94–48.64)	38.40% (34.27–42.66)	30.38% (20.32–42.02)	42.31% (31.93–53.21)
The rules on harassment were explained to everyone in the workplace.	64.52% (61.13–67.82)	68.55% (64.07–72.78)	67.51% (61.48–73.15)	56.24% (51.93–60.48)	49.97% (35.01–64.93)	60.35% (50.37–69.74)
Someone talked to the [person|people] to ask them to change their behavior.	43.48% (40.16–46.84)	43.69% (38.71–48.76)	44.15% (37.81–50.64)	45.70% (41.43–50.01)	37.72% (26.04–50.55)	53.78% (43.58–63.76)
Your work station or duties were changed to help you avoid [that person|those people].	19.97% (17.48–22.64)	18.86% (15.97–22.04)	19.92% (15.32–25.19)	19.64% (16.59–22.99)	NR (12.70–41.30)	21.51% (12.32–33.40)
The [person was moved|people were moved] or reassigned so that you did not have as much contact with them.	16.09% (13.81–18.59)	16.53% (13.83–19.52)	13.67% (9.97–18.12)	15.36% (12.58–18.49)	NR (10.32–39.29)	19.90% (10.68–32.27)
There was some official career action taken against the [person|people] for their upsetting behavior (for example, a negative evaluation/fitness report).	10.62% (9.01–12.39)	13.45% (10.66–16.65)	7.07% (4.75–10.06)	8.66% (6.35–11.47)	11.95% (7.00–18.66)	16.04% (7.58–28.33)
The [person|people] stopped their upsetting behavior.	27.48% (24.70–30.39)	28.48% (24.03–33.26)	25.27% (20.44–30.61)	30.88% (27.00–34.96)	25.45% (16.27–36.58)	34.87% (24.65–46.22)

Table B.17.b—Continued

	Total DoD	Army	Navy	Air Force	Marine Corps	Coast Guard
You were encouraged to drop the issue.	44.13% (40.58–47.72)	43.31% (38.19–48.54)	44.29% (37.50–51.23)	39.86% (35.67–44.17)	52.54% (37.76–66.99)	31.73% (21.53–43.40)
You were discouraged from filing a formal complaint.	30.25% (27.27–33.36)	34.06% (29.33–39.03)	27.59% (22.33–33.35)	25.41% (21.70–29.41)	28.68% (18.13–41.25)	29.46% (19.25–41.43)
The [person|people] who did this retaliated against you for complaining. For example, their upsetting behavior became worse or they threatened you.	31.30% (28.06–34.68)	33.70% (29.13–38.51)	32.50% (25.52–40.11)	25.76% (22.08–29.70)	25.09% (16.23–35.80)	28.40% (18.45–40.18)
Your coworkers treated you worse, avoided you, or blamed you for the problem.	31.03% (27.74–34.47)	27.09% (23.27–31.17)	31.82% (24.96–39.31)	30.32% (26.46–34.40)	NR (31.12–61.09)	24.02% (14.79–35.45)
Your supervisor punished you for bringing it up (e.g., loss of privileges, denied promotion/training, transferred to less favorable job).	20.57% (18.23–23.07)	22.63% (19.15–26.40)	19.86% (15.05–25.41)	17.73% (14.60–21.21)	17.89% (11.37–26.14)	20.62% (11.46–32.69)

NOTE: Includes estimates for active-component DoD and Coast Guard members. 95-percent confidence intervals for each estimate are included in parentheses.

For respondents who experienced sexual harassment or gender discrimination behaviors involving multiple offenders across multiple situations, this question refers to their "worst or most serious" situation.

NR = Not reportable.

Table B.17.c
Leadership actions taken in response to the disclosure among members who experienced sexual harassment or gender discrimination in the past year and disclosed it to a supervisor, leader, or official, by pay grade

SHFU10: What actions were taken in response to your [discussing the situation with a supervisor or anyone up your chain of command/reporting the situation]?

	Total	E1–E4	E5–E9	O1–O3	O4–O6
No action was taken because you asked for the discussion to be kept private.	23.86% (14.30–35.82)	23.95% (12.91–38.30)	NR (9.99–48.91)	12.97% (4.29–27.89)	NR (3.03–67.61)
You discussed the situation, but no action was taken because you chose not to give enough details about the situation.	14.53% (6.03–27.71)	12.67% (6.05–22.48)	NR (4.75–46.00)	1.47% (0.00–11.69)	NR (0.47–29.07)
The person you told took no action.	42.31% (31.93–53.21)	NR (21.53–49.65)	NR (35.63–69.39)	NR (15.79–52.70)	NR (7.14–47.66)
The rules on harassment were explained to everyone in the workplace.	60.35% (50.37–69.74)	NR (46.43–75.65)	NR (50.79–80.44)	NR (14.33–49.62)	NR (26.76–80.64)
Someone talked to the [person|people] to ask them to change their behavior.	53.78% (43.58–63.76)	NR (40.78–70.34)	NR (39.30–72.14)	NR (21.12–61.78)	NR (16.36–74.76)
Your work station or duties were changed to help you avoid [that person|those people].	21.51% (12.32–33.40)	27.67% (16.06–41.99)	NR (5.35–45.18)	7.32% (1.45–20.37)	NR (5.07–43.53)
The [person was moved|people were moved] or reassigned so that you did not have as much contact with them.	19.90% (10.68–32.27)	22.52% (11.80–36.75)	NR (7.10–46.90)	5.41% (0.72–17.72)	NR (0.21–26.96)
There was some official career action taken against the [person|people] for their upsetting behavior (for example, a negative evaluation/fitness report).	16.04% (7.58–28.33)	14.47% (7.37–24.58)	NR (4.14–44.59)	NR (4.65–32.23)	NR (0.29–31.19)
The [person|people] stopped their upsetting behavior.	34.87% (24.65–46.22)	NR (22.02–50.71)	NR (20.79–57.88)	NR (9.92–41.93)	NR (3.06–67.61)
You were encouraged to drop the issue.	31.73% (21.53–43.40)	NR (17.11–46.44)	NR (20.41–57.57)	NR (5.27–33.66)	NR (5.69–44.83)
You were discouraged from filing a formal complaint.	29.46% (19.25–41.43)	NR (11.60–41.40)	NR (19.84–57.18)	NR (7.31–38.10)	NR (4.22–41.62)

Table B.17.c—Continued

	Total	E1–E4	E5–E9	O1–O3	O4–O6
The [person|people] who did this retaliated against you for complaining. For example, their upsetting behavior became worse or they threatened you.	28.40% (18.45–40.18)	19.58% (11.02–30.87)	NR (20.26–58.18)	NR (7.20–38.62)	NR (7.80–49.81)
Your coworkers treated you worse, avoided you, or blamed you for the problem.	24.02% (14.79–35.45)	26.11% (16.14–38.27)	NR (8.83–48.21)	NR (6.25–36.26)	NR (2.82–38.09)
Your supervisor punished you for bringing it up (e.g., loss of privileges, denied promotion/ training, transferred to less favorable job).	20.62% (11.46–32.69)	12.64% (6.05–22.40)	NR (8.91–48.86)	NR (9.62–43.06)	NR (11.89–58.60)

NOTE: Includes estimates for active-component Coast Guard members. 95-percent confidence intervals for each estimate are included in parentheses.

Too few warrant officers were included in the sample to break them out as a separate pay grade. For the purposes of this table, warrant officers have been included with the E5–E9 category.

For respondents who experienced sexual harassment or gender discrimination behaviors involving multiple offenders across multiple situations, this question refers to their "worst or most serious" situation.

NR = Not reportable.

B.18. Satisfaction with the leadership response among members who experienced sexual harassment or gender discrimination in the past year and disclosed it to a supervisor, leader, or official

Table B.18.a
Satisfaction with the leadership response among members who experienced sexual harassment or gender discrimination in the past year and disclosed it to a supervisor, leader, or official, by gender

SHFU11a–f. How satisfied were/are you with the following aspects of how the discussion or report was handled? Response scale ranged from 1 (very dissatisfied) to 5 (very satisfied).

	Total	Men	Women
Availability of information about how to file a complaint	3.31 (0.15)	3.35 (0.30)	3.28 (0.09)
How you were treated by personnel handling your situation	2.99 (0.15)	3.04 (0.29)	2.94 (0.11)
The action taken by the personnel handling your situation	2.86 (0.14)	2.90 (0.28)	2.82 (0.11)
The current status of the situation	2.78 (0.13)	2.78 (0.26)	2.79 (0.09)
Amount of time it took to address your situation	2.77 (0.14)	2.62 (0.27)	2.90 (0.10)
Availability of information or updates on the status of your report or complaint	2.88 (0.14)	2.83 (0.27)	2.93 (0.09)

NOTE: Includes estimates for active-component Coast Guard members. The standard error for each estimate is included in parentheses.

For respondents who experienced sexual harassment or gender discrimination behaviors involving multiple offenders across multiple situations, this question refers to their "worst or most serious" situation.

Table B.18.b
Satisfaction with the leadership response among members who experienced sexual harassment or gender discrimination in the past year and disclosed it to a supervisor, leader, or official, by service

SHFU11a–f. How satisfied were/are you with the following aspects of how the discussion or report was handled? Response scale ranged from 1 (very dissatisfied) to 5 (very satisfied).

	Total DoD	Army	Navy	Air Force	Marine Corps	Coast Guard
Availability of information about how to file a complaint	3.19 (0.04)	3.28 (0.05)	3.15 (0.10)	3.17 (0.05)	2.98 (0.14)	3.31 (0.15)
How you were treated by personnel handling your situation	2.88 (0.04)	2.94 (0.06)	2.87 (0.08)	2.97 (0.05)	2.61 (0.12)	2.99 (0.15)
The action taken by the personnel handling your situation	2.76 (0.04)	2.78 (0.07)	2.74 (0.08)	2.82 (0.05)	2.63 (0.13)	2.86 (0.14)
The current status of the situation	2.73 (0.05)	2.73 (0.07)	2.75 (0.09)	2.74 (0.05)	2.61 (0.19)	2.78 (0.13)
Amount of time it took to address your situation	2.68 (0.04)	2.69 (0.06)	2.66 (0.09)	2.77 (0.05)	2.56 (0.12)	2.77 (0.14)
Availability of information or updates on the status of your report or complaint	2.73 (0.04)	2.74 (0.05)	2.72 (0.09)	2.85 (0.04)	2.59 (0.12)	2.88 (0.14)

NOTE: Includes estimates for active-component DoD and Coast Guard members. The standard error for each estimate is included in parentheses.

For respondents who experienced sexual harassment or gender discrimination behaviors involving multiple offenders across multiple situations, this question refers to their "worst or most serious" situation.

Table B.18.c
Satisfaction with the leadership response among members who experienced sexual harassment or gender discrimination in the past year and disclosed it to a supervisor, leader, or official, by pay grade

SHFU11a–f. How satisfied were/are you with the following aspects of how the discussion or report was handled? Response scale ranged from 1 (very dissatisfied) to 5 (very satisfied).

	Total	E1–E4	E5–E9	O1–O3	O4–O6
Availability of information about how to file a complaint	3.31 (0.15)	3.35 (0.12)	3.23 (0.28)	3.51 (0.25)	3.42 (0.25)
How you were treated by personnel handling your situation	2.99 (0.15)	3.15 (0.15)	2.84 (0.26)	3.01 (0.31)	3.16 (0.30)
The action taken by the personnel handling your situation	2.86 (0.14)	3.07 (0.16)	2.68 (0.24)	2.88 (0.34)	2.90 (0.33)
The current status of the situation	2.78 (0.13)	2.93 (0.12)	2.68 (0.24)	2.63 (0.18)	2.98 (0.50)
Amount of time it took to address your situation	2.77 (0.14)	2.80 (0.18)	2.70 (0.25)	2.93 (0.33)	2.85 (0.33)
Availability of information or updates on the status of your report or complaint	2.88 (0.14)	2.98 (0.11)	2.74 (0.25)	3.14 (0.29)	2.90 (0.18)

NOTE: Includes estimates for active-component Coast Guard members. The standard error for each estimate is included in parentheses.

Too few warrant officers were included in the sample to break them out as a separate pay grade. For the purposes of this table, warrant officers have been included with the E5–E9 category.

For respondents who experienced sexual harassment or gender discrimination behaviors involving multiple offenders across multiple situations, this question refers to their "worst or most serious" situation.

B.19. Reasons for not disclosing among members who experienced sexual harassment or gender discrimination in the past year and did not disclose it to a supervisor, leader, or official

Table B.19.a
Reasons for not disclosing among members who experienced sexual harassment or gender discrimination in the past year and did not disclose it to a supervisor, leader, or official, by gender

SHFU12: What were your reasons for not discussing it with someone above you in your chain of command and not reporting it to a person who enforces sexual harassment regulations?

	Total	Men	Women
The offensive behavior stopped on its own.	39.86% (31.71–48.45)	42.88% (29.85–56.67)	35.81% (28.72–43.39)
Someone else already reported it.	2.88% (0.90–6.75)	2.63% (0.23–10.22)	3.20% (1.14–6.97)
You thought it was not serious enough to report.	57.92% (49.71–65.82)	65.47% (52.69–76.79)	47.78% (40.12–55.51)
You did not want more people to know.	20.68% (14.85–27.57)	14.94% (6.99–26.65)	28.40% (21.92–35.61)
You did not want people to see you as weak.	30.98% (23.50–39.28)	29.46% (17.95–43.26)	33.03% (26.16–40.48)
You did not want people to think you were gay/lesbian/bisexual/transgender.	7.88% (3.52–14.81)	12.75% (5.41–24.21)	1.33% (0.14–4.93)
You wanted to forget about it and move on.	45.86% (37.52–54.37)	39.96% (27.25–53.76)	53.79% (46.08–61.37)
You did not know how to report it.	3.15% (0.94–7.57)	3.04% (0.25–12.00)	3.29% (1.38–6.54)
Someone told you not to report it.	0.81% (0.15–2.48)	0.66% (0.00–5.70)	1.02% (0.12–3.72)
You did not think anything would be done.	38.10% (30.24–46.45)	36.01% (23.89–49.59)	40.91% (33.51–48.62)
You did not think you would be believed.	11.29% (7.08–16.83)	8.50% (2.99–18.18)	15.05% (10.14–21.16)
You did not trust the process would be fair.	26.49% (19.78–34.10)	23.40% (13.49–36.03)	30.64% (23.94–37.99)
You felt partially to blame.	5.96% (3.22–9.97)	3.01% (0.34–10.69)	9.93% (5.95–15.33)
You thought other people would blame you.	16.83% (11.70–23.07)	13.42% (6.43–23.73)	21.41% (15.50–28.34)
You thought you might get in trouble for something you did.	13.66% (8.57–20.26)	13.50% (5.99–24.94)	13.86% (9.17–19.79)

Table B.19.a—Continued

	Total	Men	Women
You thought a supervisor would make too big of a deal out of it.	33.97% (26.56–42.01)	33.14% (21.85–46.05)	35.09% (27.71–43.03)
You thought you might be labeled as a troublemaker.	24.59% (17.86–32.37)	21.48% (11.64–34.50)	28.76% (21.91–36.40)
You thought it might hurt your performance evaluation/fitness report.	22.03% (16.09–28.94)	17.95% (9.57–29.39)	27.50% (21.12–34.64)
You thought it might hurt your career.	26.07% (19.66–33.33)	21.87% (12.66–33.70)	31.73% (24.96–39.12)
You did not want to hurt the person's career or family.	27.00% (19.89–35.09)	23.66% (13.21–37.12)	31.48% (24.26–39.43)
You were worried about retaliation by the person(s) who did it.	31.69% (23.98–40.22)	29.96% (18.19–44.05)	34.02% (26.69–41.96)
You were worried about retaliation by supervisor or someone in your chain of command.	22.08% (15.55–29.82)	21.86% (12.11–34.62)	22.38% (15.85–30.08)
You were worried about retaliation by your military co-workers or peers.	17.21% (11.49–24.30)	13.73% (5.92–25.74)	21.89% (15.40–29.60)
You took other actions to handle the situation.	44.43% (35.99–53.10)	48.46% (34.95–62.14)	39.00% (31.73–46.65)

NOTE: Includes estimates for active-component Coast Guard members. 95-percent confidence intervals for each estimate are included in parentheses.

For respondents who experienced sexual harassment or gender discrimination behaviors involving multiple offenders across multiple situations, this question refers to their "worst or most serious" situation.

Table B.19.b

Reasons for not disclosing among members who experienced sexual harassment or gender discrimination in the past year and did not disclose it to a supervisor, leader, or official, by service

SHFU12: What were your reasons for not discussing it with someone above you in your chain of command and not reporting it to a person who enforces sexual harassment regulations?

	Total DoD	Army	Navy	Air Force	Marine Corps	Coast Guard
The offensive behavior stopped on its own.	36.03% (33.14–39.00)	37.78% (33.08–42.67)	33.00% (27.92–38.39)	36.80% (32.22–41.56)	37.10% (27.91–47.03)	39.86% (31.71–48.45)
Someone else already reported it.	3.68% (2.44–5.31)	5.98% (3.27–9.89)	1.95% (1.16–3.07)	2.90% (1.88–4.27)	1.27% (0.31–3.39)	2.88% (0.90–6.75)
You thought it was not serious enough to report.	49.10% (45.77–52.44)	42.12% (37.16–47.19)	53.65% (46.89–60.31)	51.17% (45.84–56.48)	58.70% (48.30–68.55)	57.92% (49.71–65.82)
You did not want more people to know.	26.16% (23.30–29.17)	24.22% (20.43–28.33)	32.88% (26.47–39.80)	20.28% (17.24–23.59)	21.96% (14.88–30.49)	20.68% (14.85–27.57)
You did not want people to see you as weak.	34.16% (30.89–37.53)	31.27% (26.52–36.34)	36.51% (29.88–43.54)	32.86% (27.37–38.71)	39.37% (29.39–50.05)	30.98% (23.50–39.28)
You did not want people to think you were gay/lesbian/bisexual/transgender.	8.89% (6.49–11.81)	9.55% (6.07–14.11)	10.67% (5.62–17.93)	3.56% (2.19–5.43)	8.12% (3.37–15.92)	7.88% (3.52–14.81)
You wanted to forget about it and move on.	51.76% (48.43–55.08)	49.52% (44.37–54.68)	54.18% (47.58–60.67)	48.28% (42.87–53.72)	57.11% (46.59–67.18)	45.86% (37.52–54.37)
You did not know how to report it.	6.22% (4.37–8.55)	4.32% (2.63–6.65)	7.59% (3.60–13.74)	6.01% (4.11–8.43)	9.38% (2.96–21.06)	3.15% (0.94–7.57)
Someone told you not to report it.	3.17% (1.64–5.48)	3.85% (2.08–6.46)	3.80% (0.58–12.07)	1.34% (0.64–2.47)	1.31% (0.32–3.54)	0.81% (0.15–2.48)
You did not think anything would be done.	44.41% (41.04–47.82)	45.73% (40.59–50.95)	47.07% (40.31–53.90)	42.24% (37.16–47.45)	35.38% (25.73–45.99)	38.10% (30.24–46.45)
You did not think you would be believed.	16.87% (14.29–19.70)	20.21% (15.92–25.06)	16.75% (11.71–22.86)	13.06% (10.55–15.93)	10.13% (5.83–16.07)	11.29% (7.08–16.83)
You did not trust the process would be fair.	32.53% (29.43–35.76)	32.75% (28.07–37.70)	33.32% (27.16–39.93)	27.95% (24.20–31.94)	35.08% (24.99–46.25)	26.49% (19.78–34.10)
You felt partially to blame.	10.33% (8.80–12.03)	9.06% (6.88–11.66)	12.34% (9.25–16.01)	8.85% (7.07–10.90)	11.15% (6.32–17.84)	5.96% (3.22–9.97)
You thought other people would blame you.	21.18% (18.27–24.34)	19.83% (16.13–23.96)	26.24% (19.67–33.70)	18.04% (12.68–24.49)	16.17% (10.28–23.67)	16.83% (11.70–23.07)
You thought you might get in trouble for something you did.	15.35% (12.73–18.27)	17.69% (13.63–22.37)	16.15% (10.76–22.85)	8.39% (6.55–10.55)	13.29% (7.60–21.03)	13.66% (8.57–20.26)
You thought a supervisor would make too big of a deal out of it.	33.66% (30.42–37.03)	30.94% (26.42–35.75)	37.52% (30.92–44.48)	28.24% (22.70–34.33)	39.17% (28.64–50.50)	33.97% (26.56–42.01)

Table B.19.b—Continued

	Total DoD	Army	Navy	Air Force	Marine Corps	Coast Guard
You thought you might be labeled as a troublemaker.	28.87% (25.86–32.02)	31.65% (27.08–36.51)	27.76% (21.85–34.30)	27.62% (22.13–33.66)	23.66% (15.32–33.80)	24.59% (17.86–32.37)
You thought it might hurt your performance evaluation/fitness report.	21.86% (19.23–24.68)	21.84% (18.26–25.77)	23.71% (17.92–30.32)	20.75% (16.47–25.57)	18.34% (12.04–26.18)	22.03% (16.09–28.94)
You thought it might hurt your career.	28.42% (25.60–31.37)	29.21% (25.18–33.51)	28.78% (22.82–35.34)	29.61% (24.93–34.62)	23.34% (15.80–32.36)	26.07% (19.66–33.33)
You did not want to hurt the person's career or family.	23.94% (21.36–26.67)	21.52% (17.83–25.58)	25.64% (20.67–31.12)	26.75% (21.15–32.97)	24.53% (16.76–33.74)	27.00% (19.89–35.09)
You were worried about retaliation by the person(s) who did it.	28.57% (25.49–31.80)	30.08% (25.27–35.24)	29.26% (23.25–35.86)	29.94% (23.98–36.45)	19.94% (13.19–28.23)	31.69% (23.98–40.22)
You were worried about retaliation by supervisor or someone in your chain of command.	23.90% (21.02–26.96)	26.12% (21.75–30.87)	23.16% (17.32–29.87)	21.78% (17.42–26.67)	20.66% (13.38–29.65)	22.08% (15.55–29.82)
You were worried about retaliation by your military co-workers or peers.	24.46% (21.56–27.54)	23.08% (18.99–27.58)	27.15% (20.98–34.04)	21.32% (18.13–24.80)	25.78% (17.30–35.85)	17.21% (11.49–24.30)
You took other actions to handle the situation.	37.12% (33.85–40.49)	37.50% (32.34–42.88)	37.10% (30.87–43.66)	32.23% (28.06–36.62)	41.60% (30.97–52.84)	44.43% (35.99–53.10)

NOTE: Includes estimates for active-component DoD and Coast Guard members. 95-percent confidence intervals for each estimate are included in parentheses.

For respondents who experienced sexual harassment or gender discrimination behaviors involving multiple offenders across multiple situations, this question refers to their "worst or most serious" situation.

Table B.19.c
Reasons for not disclosing among members who experienced sexual harassment or gender discrimination in the past year and did not disclose it to a supervisor, leader, or official, by pay grade

SHFU12: What were your reasons for not discussing it with someone above you in your chain of command and not reporting it to a person who enforces sexual harassment regulations?

	Total	E1–E4	E5–E9	O1–O3	O4–O6
The offensive behavior stopped on its own.	39.86% (31.71–48.45)	40.42% (28.23–53.54)	NR (25.15–54.46)	NR (22.57–61.06)	NR (11.08–70.91)
Someone else already reported it.	2.88% (0.90–6.75)	2.81% (0.77–7.05)	4.31% (0.47–15.21)	0.34% (0.00–6.36)	NR (0.00–24.37)
You thought it was not serious enough to report.	57.92% (49.71–65.82)	69.19% (57.75–79.13)	47.01% (32.43–61.98)	NR (39.03–74.48)	NR (14.91–63.24)
You did not want more people to know.	20.68% (14.85–27.57)	20.40% (12.01–31.21)	22.30% (12.30–35.36)	NR (7.67–36.58)	NR (3.50–35.15)
You did not want people to see you as weak.	30.98% (23.50–39.28)	30.31% (18.97–43.74)	33.28% (20.53–48.11)	NR (14.27–47.39)	NR (9.10–49.34)
You did not want people to think you were gay/lesbian/bisexual/transgender.	7.88% (3.52–14.81)	4.52% (0.51–15.76)	12.33% (3.99–26.89)	NR (1.01–29.10)	NR (0.00–24.37)
You wanted to forget about it and move on.	45.86% (37.52–54.37)	38.60% (26.67–51.63)	56.78% (41.85–70.85)	NR (21.00–55.43)	NR (32.11–81.74)
You did not know how to report it.	3.15% (0.94–7.57)	6.43% (1.63–16.28)	0.18% (0.00–3.51)	1.74% (0.06–8.77)	NR (0.00–24.37)
Someone told you not to report it.	0.81% (0.15–2.48)	1.54% (0.18–5.51)	0.20% (0.00–3.56)	0.40% (0.00–6.46)	NR (0.00–24.37)
You did not think anything would be done.	38.10% (30.24–46.45)	43.83% (31.01–57.28)	36.68% (24.02–50.85)	NR (13.55–46.02)	NR (10.64–52.13)
You did not think you would be believed.	11.29% (7.08–16.83)	10.54% (4.68–19.66)	15.71% (7.74–27.10)	3.63% (0.54–11.67)	NR (1.06–25.62)
You did not trust the process would be fair.	26.49% (19.78–34.10)	26.56% (16.45–38.86)	28.01% (16.75–41.74)	NR (10.50–42.07)	NR (7.21–45.36)
You felt partially to blame.	5.96% (3.22–9.97)	5.01% (2.06–10.00)	9.04% (3.11–19.49)	2.48% (0.19–9.94)	1.31% (0.00–15.12)
You thought other people would blame you.	16.83% (11.70–23.07)	13.18% (6.77–22.37)	23.88% (13.75–36.75)	12.32% (4.52–25.30)	NR (1.46–26.40)
You thought you might get in trouble for something you did.	13.66% (8.57–20.26)	16.49% (8.54–27.57)	14.65% (6.15–27.75)	6.34% (1.44–16.81)	1.30% (0.00–15.11)
You thought a supervisor would make too big of a deal out of it.	33.97% (26.56–42.01)	37.18% (25.57–49.97)	32.23% (19.93–46.64)	NR (17.63–51.49)	NR (5.13–41.89)
You thought you might be labeled as a troublemaker.	24.59% (17.86–32.37)	27.12% (16.82–39.58)	26.09% (14.54–40.70)	NR (4.09–31.22)	NR (8.59–46.17)

Table B.19.c—Continued

	Total	E1–E4	E5–E9	O1–O3	O4–O6
You thought it might hurt your performance evaluation/fitness report.	22.03% (16.09–28.94)	19.13% (10.56–30.57)	24.57% (14.61–37.03)	NR (6.20–34.34)	NR (22.46–76.13)
You thought it might hurt your career.	26.07% (19.66–33.33)	26.95% (17.10–38.81)	27.46% (16.74–40.51)	NR (7.49–36.01)	NR (12.46–55.83)
You did not want to hurt the person's career or family.	27.00% (19.89–35.09)	32.37% (20.93–45.59)	25.81% (14.42–40.24)	NR (7.97–38.68)	NR (0.22–20.34)
You were worried about retaliation by the person(s) who did it.	31.69% (23.98–40.22)	27.98% (17.60–40.42)	NR (25.13–55.13)	NR (11.89–44.25)	NR (7.85–49.13)
You were worried about retaliation by supervisor or someone in your chain of command.	22.08% (15.55–29.82)	22.48% (12.67–35.16)	25.92% (14.65–40.13)	NR (4.12–32.26)	NR (2.13–33.81)
You were worried about retaliation by your military co-workers or peers.	17.21% (11.49–24.30)	18.40% (9.95–29.81)	18.82% (8.82–33.08)	10.97% (4.02–22.68)	NR (2.28–36.00)
You took other actions to handle the situation.	44.43% (35.99–53.10)	50.92% (37.66–64.10)	32.96% (20.32–47.70)	NR (30.15–67.70)	NR (36.11–84.03)

NOTE: Includes estimates for active-component Coast Guard members. 95-percent confidence intervals for each estimate are included in parentheses.

Too few warrant officers were included in the sample to break them out as a separate pay grade. For the purposes of this table, warrant officers have been included with the E5–E9 category.

For respondents who experienced sexual harassment or gender discrimination behaviors involving multiple offenders across multiple situations, this question refers to their "worst or most serious" situation.

NR = Not reportable.

Beliefs About Sexual Assault and Sexual Harassment Prevalence, Prevention, and Progress in the Coast Guard: Detailed Results

C.1. Perception of safety at home duty station

Table C.1.a
Perception of safety at home duty station, by gender

Longform1: To what extent do/would you feel safe from being sexually assaulted at your home duty station?

	Total	Men	Women
Very safe	86.34%	89.66%	67.52%
	(84.76–87.81)	(87.82–91.31)	(64.81–70.14)
Safe	11.24%	8.96%	24.15%
	(9.84–12.76)	(7.39–10.74)	(21.78–26.64)
Neither safe nor unsafe	2.02%	1.10%	7.21%
	(1.55–2.58)	(0.66–1.72)	(5.80–8.84)
Unsafe	0.10%	NR	0.70%
	(0.02–0.30)	(NR)	(0.32–1.33)
Very unsafe	0.30%	0.28%	0.43%
	(0.12–0.61)	(0.09–0.65)	(0.13–1.03)

NOTE: Includes estimates for active-component Coast Guard members. 95-percent confidence intervals for each estimate are included in parentheses.

NR = Not reportable.

Table C.1.b
Perception of safety at home duty station, by service

Longform1: To what extent do/would you feel safe from being sexually assaulted at your home duty station?

	Total DoD	Army	Navy	Air Force	Marine Corps	Coast Guard
Very safe	73.81%	70.37%	73.96%	81.66%	69.85%	86.34%
	(72.47–75.12)	(68.15–72.51)	(70.98–76.78)	(80.40–82.87)	(64.73–74.63)	(84.76–87.81)
Safe	19.65%	22.31%	20.30%	14.15%	20.48%	11.24%
	(18.46–20.87)	(20.33–24.39)	(17.71–23.08)	(13.13–15.23)	(16.29–25.19)	(9.84–12.76)
Neither safe nor unsafe	5.06%	5.40%	4.88%	3.67%	6.73%	2.02%
	(4.47–5.71)	(4.49–6.44)	(3.86–6.08)	(2.98–4.48)	(4.32–9.91)	(1.55–2.58)
Unsafe	0.75%	1.10%	0.42%	0.17%	1.33%	0.10%
	(0.47–1.15)	(0.62–1.82)	(0.19–0.81)	(0.08–0.31)	(0.24–4.10)	(0.02–0.30)
Very unsafe	0.73%	0.82%	0.44%	0.34%	1.61%	0.30%
	(0.48–1.07)	(0.52–1.22)	(0.19–0.87)	(0.19–0.56)	(0.40–4.28)	(0.12–0.61)

NOTE: Includes estimates for active-component DoD and Coast Guard members. 95-percent confidence intervals for each estimate are included in parentheses.

Table C.1.c
Perception of safety at home duty station, by pay grade

Longform1: To what extent do/would you feel safe from being sexually assaulted at your home duty station?

	Total	E1–E4	E5–E9	O1–O3	O4–O6
Very safe	86.34%	81.75%	87.69%	88.31%	93.61%
	(84.76–87.81)	(78.42–84.76)	(85.44–89.71)	(84.69–91.33)	(89.21–96.61)
Safe	11.24%	14.89%	9.96%	10.54%	5.75%
	(9.84–12.76)	(12.06–18.10)	(8.05–12.14)	(7.61–14.11)	(2.91–10.04)
Neither safe nor unsafe	2.02%	2.96%	1.91%	0.98%	0.12%
	(1.55–2.58)	(1.95–4.30)	(1.27–2.75)	(0.28–2.46)	(0.00–1.55)
Unsafe	0.10%	0.15%	0.10%	0.08%	0.00%
	(0.02–0.30)	(0.01–0.69)	(0.01–0.43)	(0.00–1.03)	(0.00–1.68)
Very unsafe	0.30%	0.24%	0.34%	0.08%	0.52%
	(0.12–0.61)	(0.02–1.06)	(0.10–0.83)	(0.00–1.03)	(0.01–2.86)

NOTE: Includes estimates for active-component Coast Guard members. 95-percent confidence intervals for each estimate are included in parentheses.

Too few warrant officers were included in the sample to break them out as a separate pay grade. For the purposes of this table, warrant officers have been included with the E5–E9 category.

C.2. Perception of safety away from home duty station

Table C.2.a
Perception of safety away from home duty station, by gender

Longform2: To what extent do/would you feel safe from being sexually assaulted during military operations, training, or exercises away from your home duty station?

	Total	Men	Women
Very safe	79.88% (78.11–81.57)	85.61% (83.57–87.49)	47.61% (44.78–50.45)
Safe	15.86% (14.30–17.52)	12.25% (10.50–14.18)	36.20% (33.53–38.94)
Neither safe nor unsafe	3.58% (2.90–4.36)	1.87% (1.19–2.79)	13.18% (11.33–15.23)
Unsafe	0.38% (0.19–0.67)	0.10% (0.01–0.40)	1.94% (1.23–2.89)
Very unsafe	0.30% (0.14–0.57)	0.17% (0.03–0.49)	1.07% (0.53–1.91)

NOTE: Includes estimates for active-component Coast Guard members. 95-percent confidence intervals for each estimate are included in parentheses.

Table C.2.b
Perception of safety away from home duty station, by service

Longform2: To what extent do/would you feel safe from being sexually assaulted during military operations, training, or exercises away from your home duty station?

	Total DoD	Army	Navy	Air Force	Marine Corps	Coast Guard
Very safe	68.87% (67.54–70.17)	66.93% (64.73–69.07)	66.86% (63.65–69.96)	73.06% (71.68–74.42)	70.50% (65.79–74.92)	79.88% (78.11–81.57)
Safe	22.18% (21.03–23.36)	23.73% (21.79–25.75)	24.07% (21.35–26.96)	19.29% (18.14–20.48)	19.62% (16.00–23.67)	15.86% (14.30–17.52)
Neither safe nor unsafe	7.20% (6.51–7.93)	7.11% (6.08–8.26)	7.76% (6.39–9.31)	6.49% (5.68–7.38)	7.65% (5.00–11.11)	3.58% (2.90–4.36)
Unsafe	1.08% (0.87–1.33)	1.58% (1.12–2.17)	0.81% (0.50–1.24)	0.86% (0.64–1.12)	0.55% (0.24–1.09)	0.38% (0.19–0.67)
Very unsafe	0.68% (0.43–1.02)	0.65% (0.38–1.02)	0.50% (0.23–0.96)	0.29% (0.16–0.49)	1.67% (0.43–4.33)	0.30% (0.14–0.57)

NOTE: Includes estimates for active-component DoD and Coast Guard members. 95-percent confidence intervals for each estimate are included in parentheses.

Table C.2.c

Perception of safety away from home duty station, by pay grade

Longform2: To what extent do/would you feel safe from being sexually assaulted during military operations, training, or exercises away from your home duty station?

	Total	E1–E4	E5–E9	O1–O3	O4–O6
Very safe	79.88%	75.15%	81.90%	78.84%	86.99%
	(78.11–81.57)	(71.55–78.50)	(79.43–84.19)	(73.35–83.66)	(82.09–90.96)
Safe	15.86%	19.50%	14.01%	17.17%	11.87%
	(14.30–17.52)	(16.39–22.90)	(11.93–16.30)	(12.53–22.67)	(8.06–16.66)
Neither safe nor unsafe	3.58%	4.45%	3.47%	3.64%	0.47%
	(2.90–4.36)	(3.19–6.02)	(2.47–4.73)	(2.08–5.90)	(0.02–2.18)
Unsafe	0.38%	0.45%	0.39%	0.26%	0.15%
	(0.19–0.67)	(0.12–1.15)	(0.14–0.87)	(0.01–1.34)	(0.00–1.61)
Very unsafe	0.30%	0.46%	0.22%	0.08%	0.52%
	(0.14–0.57)	(0.13–1.16)	(0.05–0.62)	(0.00–1.03)	(0.01–2.85)

NOTE: Includes estimates for active-component Coast Guard members. 95-percent confidence intervals for each estimate are included in parentheses.

Too few warrant officers were included in the sample to break them out as a separate pay grade. For the purposes of this table, warrant officers have been included with the E5–E9 category.

C.3. Perception of how common sexual harassment is in the military

Table C.3.a
Perception of how common sexual harassment is in the military, by gender

Longform3: How common is sexual harassment in the military?

	Total	Men	Women
Very common	7.65% (6.61–8.80)	5.74% (4.61–7.06)	18.40% (16.20–20.76)
Common	35.97% (33.85–38.13)	33.10% (30.66–35.60)	52.15% (49.30–54.99)
Rare	43.79% (41.61–45.99)	46.88% (44.35–49.43)	26.37% (23.96–28.88)
Very rare	12.59% (11.15–14.15)	14.28% (12.59–16.10)	3.08% (2.04–4.47)

NOTE: Includes estimates for active-component Coast Guard members.
95-percent confidence intervals for each estimate are included in parentheses.

Table C.3.b
Perception of how common sexual harassment is in the military, by service

Longform3: How common is sexual harassment in the military?

	Total DoD	Army	Navy	Air Force	Marine Corps	Coast Guard
Very common	11.58% (10.64–12.56)	13.90% (12.27–15.66)	10.86% (8.98–12.99)	7.83% (7.10–8.62)	12.65% (9.48–16.41)	7.65% (6.61–8.80)
Common	38.00% (36.59–39.42)	41.47% (39.17–43.81)	38.06% (34.70–41.51)	33.29% (31.78–34.81)	36.25% (31.58–41.11)	35.97% (33.85–38.13)
Rare	37.66% (36.20–39.13)	33.12% (31.00–35.29)	40.87% (36.92–44.90)	42.32% (40.73–43.91)	36.87% (32.55–41.35)	43.79% (41.61–45.99)
Very rare	12.77% (11.79–13.80)	11.51% (9.73–13.49)	10.21% (8.72–11.86)	16.56% (15.30–17.88)	14.24% (10.87–18.18)	12.59% (11.15–14.15)

NOTE: Includes estimates for active-component DoD and Coast Guard members. 95-percent confidence intervals for each estimate are included in parentheses.

Table C.3.c

Perception of how common sexual harassment is in the military, by pay grade

Longform3: How common is sexual harassment in the military?

	Total	E1–E4	E5–E9	O1–O3	O4–O6
Very common	7.65% (6.61–8.80)	11.64% (9.27–14.38)	7.04% (5.72–8.55)	3.14% (1.69–5.28)	0.87% (0.14–2.83)
Common	35.97% (33.85–38.13)	43.13% (38.88–47.46)	35.10% (32.23–38.06)	31.27% (25.55–37.45)	17.44% (12.81–22.91)
Rare	43.79% (41.61–45.99)	35.51% (31.36–39.84)	45.24% (42.26–48.24)	50.76% (44.44–57.07)	59.78% (51.95–67.26)
Very rare	12.59% (11.15–14.15)	9.71% (7.19–12.75)	12.62% (10.76–14.67)	14.83% (10.58–19.98)	21.91% (15.02–30.16)

NOTE: Includes estimates for active-component Coast Guard members. 95-percent confidence intervals for each estimate are included in parentheses.

Too few warrant officers were included in the sample to break them out as a separate pay grade. For the purposes of this table, warrant officers have been included with the E5–E9 category.

C.4. Perception of how common discrimination against women is in the military

Table C.4.a
Perception of how common discrimination against women is in the military, by gender

Longform4: How common is discrimination against women in the military?

	Total	Men	Women
Very common	5.37% (4.55–6.28)	3.28% (2.43–4.32)	17.11% (15.01–19.38)
Common	26.83% (24.92–28.80)	23.56% (21.38–25.84)	45.25% (42.43–48.09)
Rare	49.37% (47.16–51.58)	52.14% (49.58–54.69)	33.74% (31.12–36.44)
Very rare	18.44% (16.71–20.27)	21.02% (19.00–23.15)	3.89% (2.76–5.31)

NOTE: Includes estimates for active-component Coast Guard members. 95-percent confidence intervals for each estimate are included in parentheses.

Table C.4.b
Perception of how common discrimination against women is in the military, by service

Longform4: How common is discrimination against women in the military?

	Total DoD	Army	Navy	Air Force	Marine Corps	Coast Guard
Very common	10.06% (9.07–11.12)	10.79% (9.29–12.43)	8.87% (7.09–10.91)	4.97% (4.42–5.58)	18.47% (14.16–23.44)	5.37% (4.55–6.28)
Common	29.56% (28.25–30.89)	33.33% (31.22–35.50)	27.55% (24.62–30.64)	23.77% (22.49–25.10)	32.23% (27.56–37.18)	26.83% (24.92–28.80)
Rare	42.11% (40.67–43.57)	41.00% (38.71–43.33)	44.72% (40.98–48.50)	48.50% (46.88–50.12)	30.23% (26.66–33.99)	49.37% (47.16–51.58)
Very rare	18.27% (17.05–19.54)	14.87% (12.96–16.95)	18.86% (15.67–22.39)	22.75% (21.36–24.19)	19.07% (15.72–22.79)	18.44% (16.71–20.27)

NOTE: Includes estimates for active-component DoD and Coast Guard members. 95-percent confidence intervals for each estimate are included in parentheses.

Table C.4.c
Perception of how common discrimination against women is in the military, by pay grade

Longform4: How common is discrimination against women in the military?

	Total	E1–E4	E5–E9	O1–O3	O4–O6
Very common	5.37% (4.55–6.28)	7.88% (6.10–9.98)	4.53% (3.48–5.78)	4.28% (2.56–6.68)	2.06% (0.73–4.52)
Common	26.83% (24.92–28.80)	31.11% (27.30–35.11)	26.05% (23.42–28.81)	24.59% (19.93–29.74)	17.09% (12.22–22.92)
Rare	49.37% (47.16–51.58)	45.45% (41.11–49.84)	50.40% (47.40–53.40)	53.19% (46.92–59.38)	53.55% (45.86–61.11)
Very rare	18.44% (16.71–20.27)	15.56% (12.38–19.20)	19.02% (16.73–21.48)	17.94% (13.30–23.39)	27.31% (20.12–35.49)

NOTE: Includes estimates for active-component Coast Guard members. 95-percent confidence intervals for each estimate are included in parentheses.

Too few warrant officers were included in the sample to break them out as a separate pay grade. For the purposes of this table, warrant officers have been included with the E5–E9 category.

C.5. Perceived likelihood that sexual harassment in the military would be reported

Table C.5.a
Perceived likelihood that sexual harassment in the military would be reported, by gender

Longform5: In the military, how likely is it that an instance of sexual harassment would be reported?

	Total	Men	Women
Very likely	11.51% (10.13–13.01)	13.03% (11.41–14.78)	2.96% (2.08–4.07)
Likely	40.01% (37.86–42.19)	42.51% (40.01–45.03)	25.92% (23.49–28.46)
Neither likely nor unlikely	27.05% (25.10–29.06)	26.23% (23.99–28.57)	31.66% (29.06–34.35)
Unlikely	18.58% (16.99–20.26)	15.68% (13.88–17.61)	34.97% (32.30–37.71)
Very unlikely	2.85% (2.15–3.68)	2.55% (1.78–3.54)	4.49% (3.35–5.89)

NOTE: Includes estimates for active-component Coast Guard members. 95-percent confidence intervals for each estimate are included in parentheses.

Table C.5.b
Perceived likelihood that sexual harassment in the military would be reported, by service

Longform5: In the military, how likely is it that an instance of sexual harassment would be reported?

	Total DoD	Army	Navy	Air Force	Marine Corps	Coast Guard
Very likely	13.33% (12.37–14.34)	15.79% (14.05–17.65)	11.25% (9.12–13.69)	11.70% (10.65–12.81)	12.85% (10.30–15.76)	11.51% (10.13–13.01)
Likely	36.55% (35.20–37.92)	36.86% (34.72–39.04)	36.41% (33.17–39.75)	34.83% (33.30–36.39)	38.78% (34.17–43.54)	40.01% (37.86–42.19)
Neither likely nor unlikely	28.02% (26.74–29.33)	27.81% (25.71–29.99)	26.49% (23.77–29.34)	29.48% (28.04–30.95)	28.74% (24.18–33.65)	27.05% (25.10–29.06)
Unlikely	18.81% (17.43–20.26)	16.54% (14.69–18.52)	22.37% (18.16–27.03)	20.88% (19.59–22.21)	15.61% (12.38–19.30)	18.58% (16.99–20.26)
Very unlikely	3.29% (2.73–3.93)	3.00% (1.85–4.59)	3.48% (2.64–4.50)	3.11% (2.61–3.68)	4.02% (2.74–5.68)	2.85% (2.15–3.68)

NOTE: Includes estimates for active-component DoD and Coast Guard members. 95-percent confidence intervals for each estimate are included in parentheses.

Table C.5.c
Perceived likelihood that sexual harassment in the military would be reported, by pay grade

Longform5: In the military, how likely is it that an instance of sexual harassment would be reported?

	Total	E1–E4	E5–E9	O1–O3	O4–O6
Very likely	11.51% (10.13–13.01)	9.52% (7.03–12.52)	13.35% (11.39–15.52)	8.21% (5.06–12.44)	11.32% (7.26–16.57)
Likely	40.01% (37.86–42.19)	31.01% (27.01–35.22)	44.93% (41.97–47.91)	39.54% (33.26–46.09)	43.95% (36.33–51.77)
Neither likely nor unlikely	27.05% (25.10–29.06)	29.59% (25.79–33.61)	25.00% (22.34–27.80)	29.45% (24.14–35.22)	27.65% (21.61–34.36)
Unlikely	18.58% (16.99–20.26)	25.09% (21.56–28.89)	14.68% (12.76–16.77)	20.61% (16.42–25.32)	15.96% (11.53–21.27)
Very unlikely	2.85% (2.15–3.68)	4.80% (3.06–7.11)	2.04% (1.34–2.98)	2.19% (1.02–4.07)	1.13% (0.24–3.20)

NOTE: Includes estimates for active-component Coast Guard members. 95-percent confidence intervals for each estimate are included in parentheses.

Too few warrant officers were included in the sample to break them out as a separate pay grade. For the purposes of this table, warrant officers have been included with the E5–E9 category.

C.6. Perceived likelihood that reports of sexual harassment in the military would be acted upon

Table C.6.a
Perceived likelihood that reports of sexual harassment in the military would be acted upon, by gender

Longform6: In the military, how likely is it that something would be done to try to stop the sexual harassment after it is reported?

	Total	Men	Women
Very likely	51.33% (49.12–53.53)	55.45% (52.90–57.99)	28.03% (25.51–30.65)
Likely	34.17% (32.13–36.25)	31.96% (29.62–34.36)	46.66% (43.83–49.50)
Neither likely nor unlikely	9.96% (8.63–11.41)	8.73% (7.23–10.43)	16.87% (14.78–19.13)
Unlikely	3.31% (2.60–4.16)	2.66% (1.87–3.67)	6.99% (5.67–8.51)
Very unlikely	1.23% (0.82–1.78)	1.19% (0.73–1.85)	1.45% (0.82–2.36)

NOTE: Includes estimates for active-component Coast Guard members. 95-percent confidence intervals for each estimate are included in parentheses.

Table C.6.b
Perceived likelihood that reports of sexual harassment in the military would be acted upon, by service

Longform6: In the military, how likely is it that something would be done to try to stop the sexual harassment after it is reported?

	Total DoD	Army	Navy	Air Force	Marine Corps	Coast Guard
Very likely	47.42% (45.94–48.90)	43.93% (41.62–46.26)	46.87% (43.15–50.62)	49.86% (48.25–51.48)	53.75% (48.87–58.59)	51.33% (49.12–53.53)
Likely	33.80% (32.33–35.29)	35.57% (33.22–37.98)	35.39% (31.70–39.21)	31.56% (30.14–33.01)	30.03% (25.40–34.97)	34.17% (32.13–36.25)
Neither likely nor unlikely	12.70% (11.80–13.64)	13.90% (12.29–15.64)	11.50% (9.73–13.47)	12.57% (11.58–13.62)	11.69% (8.87–15.02)	9.96% (8.63–11.41)
Unlikely	4.39% (3.74–5.11)	4.79% (3.83–5.91)	4.31% (2.53–6.82)	4.67% (3.87–5.59)	2.96% (1.91–4.36)	3.31% (2.60–4.16)
Very unlikely	1.69% (1.38–2.04)	1.80% (1.24–2.53)	1.93% (1.28–2.79)	1.33% (1.00–1.73)	1.57% (0.95–2.44)	1.23% (0.82–1.78)

NOTE: Includes estimates for active-component DoD and Coast Guard members. 95-percent confidence intervals for each estimate are included in parentheses.

Table C.6.c
Perceived likelihood that reports of sexual harassment in the military would be acted upon, by pay grade

Longform6: In the military, how likely is it that something would be done to try to stop the sexual harassment after it is reported?

	Total	E1–E4	E5–E9	O1–O3	O4–O6
Very likely	51.33%	41.42%	55.06%	50.27%	68.29%
	(49.12–53.53)	(37.11–45.84)	(52.06–58.03)	(43.99–56.53)	(61.57–74.49)
Likely	34.17%	36.62%	32.55%	39.48%	28.06%
	(32.13–36.25)	(32.56–40.83)	(29.81–35.38)	(33.59–45.61)	(22.16–34.58)
Neither likely nor unlikely	9.96%	15.30%	8.27%	7.50%	2.59%
	(8.63–11.41)	(12.42–18.56)	(6.53–10.31)	(5.09–10.59)	(1.06–5.22)
Unlikely	3.31%	5.33%	2.64%	2.21%	1.06%
	(2.60–4.16)	(3.57–7.63)	(1.86–3.63)	(1.04–4.11)	(0.21–3.11)
Very unlikely	1.23%	1.32%	1.48%	0.54%	0.00%
	(0.82–1.78)	(0.56–2.62)	(0.89–2.32)	(0.07–1.90)	(0.00–1.69)

NOTE: Includes estimates for active-component Coast Guard members. 95-percent confidence intervals for each estimate are included in parentheses.

Too few warrant officers were included in the sample to break them out as a separate pay grade. For the purposes of this table, warrant officers have been included with the E5–E9 category.

C.7. Perceived likelihood that sexual assault in the military would be reported

Table C.7.a
Perceived likelihood that sexual assault in the military would be reported, by gender

Longform7: In the military, how likely is it that an instance of sexual assault would be reported?

	Total	Men	Women
Very likely	19.06% (17.37–20.84)	21.43% (19.45–23.51)	5.63% (4.32–7.20)
Likely	43.49% (41.31–45.70)	45.37% (42.84–47.91)	32.88% (30.27–35.58)
Neither likely nor unlikely	23.54% (21.70–25.45)	21.58% (19.48–23.80)	34.63% (31.94–37.39)
Unlikely	11.83% (10.53–13.23)	9.72% (8.27–11.33)	23.77% (21.43–26.23)
Very unlikely	2.08% (1.46–2.86)	1.90% (1.21–2.83)	3.08% (2.15–4.28)

NOTE: Includes estimates for active-component Coast Guard members. 95-percent confidence intervals for each estimate are included in parentheses.

Table C.7.b
Perceived likelihood that sexual assault in the military would be reported, by service

Longform7: In the military, how likely is it that an instance of sexual assault would be reported?

	Total DoD	Army	Navy	Air Force	Marine Corps	Coast Guard
Very likely	21.60% (20.44–22.80)	25.01% (23.02–27.07)	18.02% (15.63–20.61)	17.65% (16.42–18.93)	25.00% (20.95–29.40)	19.06% (17.37–20.84)
Likely	42.03% (40.59–43.49)	41.58% (39.24–43.95)	44.42% (40.82–48.08)	40.53% (38.95–42.12)	41.71% (37.03–46.50)	43.49% (41.31–45.70)
Neither likely nor unlikely	21.89% (20.80–23.01)	20.88% (19.11–22.73)	20.81% (18.31–23.49)	26.22% (24.81–27.68)	19.22% (15.88–22.93)	23.54% (21.70–25.45)
Unlikely	12.72% (11.42–14.11)	10.87% (9.26–12.65)	14.86% (10.87–19.63)	13.52% (12.44–14.65)	12.81% (9.47–16.79)	11.83% (10.53–13.23)
Very unlikely	1.76% (1.46–2.10)	1.66% (1.11–2.39)	1.89% (1.27–2.69)	2.08% (1.67–2.56)	1.27% (0.74–2.01)	2.08% (1.46–2.86)

NOTE: Includes estimates for active-component DoD and Coast Guard members. 95-percent confidence intervals for each estimate are included in parentheses.

Table C.7.c
Perceived likelihood that sexual assault in the military would be reported, by pay grade

Longform7: In the military, how likely is it that an instance of sexual assault would be reported?

	Total	E1–E4	E5–E9	O1–O3	O4–O6
Very likely	19.06%	14.69%	22.59%	13.45%	20.03%
	(17.37–20.84)	(11.72–18.08)	(20.15–25.18)	(9.54–18.23)	(14.67–26.32)
Likely	43.49%	37.96%	45.57%	48.24%	45.72%
	(41.31–45.70)	(33.74–42.32)	(42.61–48.56)	(41.92–54.59)	(38.13–53.46)
Neither likely nor unlikely	23.54%	27.01%	20.93%	25.06%	25.40%
	(21.70–25.45)	(23.35–30.92)	(18.46–23.58)	(20.32–30.29)	(19.60–31.92)
Unlikely	11.83%	16.30%	9.65%	11.63%	8.59%
	(10.53–13.23)	(13.40–19.53)	(8.06–11.44)	(8.59–15.29)	(5.47–12.71)
Very unlikely	2.08%	4.04%	1.25%	1.61%	0.26%
	(1.46–2.86)	(2.41–6.32)	(0.72–2.01)	(0.58–3.54)	(0.00–1.83)

NOTE: Includes estimates for active-component Coast Guard members. 95-percent confidence intervals for each estimate are included in parentheses.

Too few warrant officers were included in the sample to break them out as a separate pay grade. For the purposes of this table, warrant officers have been included with the E5–E9 category.

C.8. Perceived likelihood that a reported sexual assault in the military would be investigated

Table C.8.a
Perceived likelihood that a reported sexual assault in the military would be investigated, by gender

Longform8: In the military, how likely is it that there would be an investigation after an unrestricted report of sexual assault?

	Total	Men	Women
Very likely	64.03% (61.90–66.12)	66.64% (64.17–69.04)	49.33% (46.49–52.17)
Likely	26.19% (24.30–28.16)	24.55% (22.37–26.82)	35.49% (32.82–38.23)
Neither likely nor unlikely	7.82% (6.69–9.07)	7.01% (5.74–8.47)	12.36% (10.57–14.33)
Unlikely	1.21% (0.77–1.80)	1.05% (0.57–1.77)	2.09% (1.35–3.08)
Very unlikely	0.74% (0.42–1.22)	0.75% (0.38–1.31)	0.73% (0.31–1.47)

NOTE: Includes estimates for active-component Coast Guard members. 95-percent confidence intervals for each estimate are included in parentheses.

Table C.8.b
Perceived likelihood that a reported sexual assault in the military would be investigated, by service

Longform8: In the military, how likely is it that there would be an investigation after an unrestricted report of sexual assault?

	Total DoD	Army	Navy	Air Force	Marine Corps	Coast Guard
Very likely	58.58% (57.07–60.09)	53.39% (50.98–55.78)	59.76% (55.81–63.62)	61.54% (59.96–63.10)	65.82% (61.06–70.35)	64.03% (61.90–66.12)
Likely	28.05% (26.63–29.51)	30.25% (27.98–32.61)	28.99% (25.21–33.00)	26.36% (25.00–27.76)	23.29% (19.37–27.57)	26.19% (24.30–28.16)
Neither likely nor unlikely	9.96% (9.14–10.83)	12.47% (10.90–14.17)	7.60% (6.34–9.01)	9.39% (8.50–10.34)	8.06% (5.42–11.43)	7.82% (6.69–9.07)
Unlikely	2.37% (1.83–3.02)	2.50% (1.92–3.21)	2.91% (1.26–5.65)	1.73% (1.39–2.13)	2.15% (0.97–4.08)	1.21% (0.77–1.80)
Very unlikely	1.04% (0.75–1.39)	1.39% (0.85–2.14)	0.74% (0.35–1.37)	0.98% (0.46–1.81)	0.69% (0.24–1.53)	0.74% (0.42–1.22)

NOTE: Includes estimates for active-component DoD and Coast Guard members. 95-percent confidence intervals for each estimate are included in parentheses.

Table C.8.c
Perceived likelihood that a reported sexual assault in the military would be investigated, by pay grade

Longform8: In the military, how likely is it that there would be an investigation after an unrestricted report of sexual assault?

	Total	E1–E4	E5–E9	O1–O3	O4–O6
Very likely	64.03% (61.90–66.12)	50.86% (46.49–55.22)	67.60% (64.80–70.30)	72.09% (66.59–77.13)	83.61% (78.21–88.13)
Likely	26.19% (24.30–28.16)	33.86% (29.80–38.09)	23.84% (21.41–26.40)	22.87% (18.19–28.11)	15.01% (10.64–20.30)
Neither likely nor unlikely	7.82% (6.69–9.07)	11.99% (9.35–15.06)	7.07% (5.69–8.66)	3.92% (2.22–6.38)	0.73% (0.09–2.60)
Unlikely	1.21% (0.77–1.80)	2.19% (1.09–3.91)	0.77% (0.37–1.42)	0.94% (0.24–2.47)	0.54% (0.01–2.96)
Very unlikely	0.74% (0.42–1.22)	1.10% (0.41–2.38)	0.73% (0.32–1.40)	0.18% (0.00–1.22)	0.12% (0.00–1.56)

NOTE: Includes estimates for active-component Coast Guard members. 95-percent confidence intervals for each estimate are included in parentheses.

Too few warrant officers were included in the sample to break them out as a separate pay grade. For the purposes of this table, warrant officers have been included with the E5–E9 category.

C.9. Perceived likelihood that someone who committed a sexual assault in the military would be punished

Table C.9.a
Perceived likelihood that someone who committed a sexual assault in the military would be punished, by gender

Longform9: In the military, how likely is it that a person who sexually assaulted someone would be held accountable or punished?

	Total	Men	Women
Very likely	58.25% (56.09–60.39)	63.20% (60.70–65.64)	30.34% (27.75–33.02)
Likely	28.01% (26.09–29.99)	25.65% (23.45–27.94)	41.32% (38.54–44.14)
Neither likely nor unlikely	9.40% (8.22–10.68)	7.83% (6.52–9.31)	18.23% (16.11–20.49)
Unlikely	2.99% (2.36–3.74)	2.26% (1.57–3.13)	7.12% (5.75–8.71)
Very unlikely	1.36% (0.95–1.89)	1.07% (0.63–1.70)	2.99% (2.11–4.12)

NOTE: Includes estimates for active-component Coast Guard members. 95-percent confidence intervals for each estimate are included in parentheses.

Table C.9.b
Perceived likelihood that someone who committed a sexual assault in the military would be punished, by service

Longform9: In the military, how likely is it that a person who sexually assaulted someone would be held accountable or punished?

	Total DoD	Army	Navy	Air Force	Marine Corps	Coast Guard
Very likely	53.27% (51.78–54.75)	47.53% (45.15–49.93)	55.19% (51.36–58.97)	52.67% (51.06–54.29)	66.59% (62.03–70.93)	58.25% (56.09–60.39)
Likely	28.00% (26.78–29.25)	30.75% (28.65–32.92)	27.07% (24.21–30.07)	29.64% (28.21–31.10)	19.38% (16.10–23.01)	28.01% (26.09–29.99)
Neither likely nor unlikely	12.21% (11.09–13.40)	14.51% (12.86–16.28)	11.17% (8.00–15.04)	11.23% (10.27–12.24)	9.34% (6.53–12.84)	9.40% (8.22–10.68)
Unlikely	4.47% (3.84–5.15)	4.85% (4.08–5.72)	4.70% (2.88–7.19)	4.26% (3.72–4.86)	3.37% (1.79–5.72)	2.99% (2.36–3.74)
Very unlikely	2.06% (1.68–2.49)	2.36% (1.64–3.26)	1.88% (1.20–2.80)	2.19% (1.56–2.99)	1.32% (0.75–2.17)	1.36% (0.95–1.89)

NOTE: Includes estimates for active-component DoD and Coast Guard members. 95-percent confidence intervals for each estimate are included in parentheses.

Table C.9.c
Perceived likelihood that someone who committed a sexual assault in the military would be punished, by pay grade

Longform9: In the military, how likely is it that a person who sexually assaulted someone would be held accountable or punished?

	Total	E1–E4	E5–E9	O1–O3	O4–O6
Very likely	58.25% (56.09–60.39)	49.65% (45.28–54.02)	63.11% (60.24–65.92)	55.80% (49.65–61.81)	63.27% (56.23–69.91)
Likely	28.01% (26.09–29.99)	30.73% (26.84–34.85)	25.77% (23.24–28.43)	30.74% (25.45–36.42)	28.73% (22.71–35.37)
Neither likely nor unlikely	9.40% (8.22–10.68)	13.31% (10.61–16.41)	7.56% (6.20–9.11)	9.32% (6.51–12.82)	6.03% (3.53–9.51)
Unlikely	2.99% (2.36–3.74)	4.23% (2.80–6.10)	2.31% (1.59–3.24)	3.69% (2.07–6.04)	1.62% (0.48–3.93)
Very unlikely	1.36% (0.95–1.89)	2.08% (1.15–3.45)	1.25% (0.73–1.97)	0.46% (0.05–1.67)	0.35% (0.01–1.98)

NOTE: Includes estimates for active-component Coast Guard members. 95-percent confidence intervals for each estimate are included in parentheses.

Too few warrant officers were included in the sample to break them out as a separate pay grade. For the purposes of this table, warrant officers have been included with the E5–E9 category.

C.10. Perceived likelihood of respondent taking specific actions related to sexual assault or harassment

Table C.10.a
Perceived likelihood of respondent taking specific actions related to sexual assault or harassment, by gender

Longform10a–g: How likely would you be to...

	Total	Men	Women
Encourage someone who has experienced sexual harassment to tell a supervisor?			
Very Likely	79.94% (78.15–81.64)	81.35% (79.28–83.30)	71.98% (69.34–74.51)
Likely	14.48% (12.98–16.08)	13.76% (12.05–15.62)	18.52% (16.32–20.88)
Neither Likely nor Unlikely	3.75% (2.98–4.64)	3.22% (2.36–4.27)	6.74% (5.41–8.27)
Unlikely	0.65% (0.37–1.05)	0.50% (0.21–1.01)	1.45% (0.87–2.25)
Very Unlikely	1.19% (0.81–1.69)	1.17% (0.72–1.79)	1.32% (0.73–2.18)
Encourage someone who has experienced sexual assault to seek counseling?			
Very Likely	83.49% (81.69–85.18)	83.65% (81.55–85.60)	82.59% (80.34–84.69)
Likely	12.54% (11.02–14.18)	12.24% (10.49–14.16)	14.22% (12.31–16.31)
Neither Likely nor Unlikely	2.78% (2.06–3.67)	2.81% (1.98–3.85)	2.65% (1.79–3.77)
Unlikely	0.24% (0.08–0.58)	0.25% (0.06–0.66)	0.22% (0.04–0.65)
Very Unlikely	0.95% (0.59–1.43)	1.06% (0.63–1.66)	0.31% (0.08–0.84)
Encourage someone who has experienced sexual assault to report it?			
Very Likely	84.74% (83.10–86.29)	85.82% (83.91–87.59)	78.65% (76.21–80.94)
Likely	11.01% (9.67–12.47)	10.18% (8.66–11.87)	15.69% (13.66–17.89)
Neither Likely nor Unlikely	2.85% (2.16–3.70)	2.58% (1.79–3.58)	4.41% (3.35–5.68)
Unlikely	0.29% (0.11–0.64)	0.30% (0.09–0.72)	0.24% (0.05–0.69)
Very Unlikely	1.10% (0.73–1.60)	1.12% (0.68–1.73)	1.01% (0.47–1.91)

Table C.10.a—Continued

	Total	Men	Women
Tell a supervisor about sexual harassment if it happened to you?			
Very Likely	66.63% (64.54–68.66)	69.11% (66.70–71.44)	52.70% (49.85–55.53)
Likely	19.76% (18.06–21.55)	19.01% (17.05–21.09)	24.00% (21.62–26.51)
Neither Likely nor Unlikely	8.10% (6.97–9.34)	7.19% (5.92–8.63)	13.21% (11.36–15.24)
Unlikely	3.33% (2.61–4.18)	2.67% (1.87–3.68)	7.02% (5.67–8.57)
Very Unlikely	2.18% (1.65–2.83)	2.03% (1.41–2.81)	3.07% (2.17–4.22)
Report a sexual assault if it happened to you?			
Very Likely	73.22% (71.24–75.14)	74.97% (72.68–77.17)	63.38% (60.61–66.08)
Likely	16.44% (14.85–18.14)	15.78% (13.95–17.75)	20.17% (17.97–22.51)
Neither Likely nor Unlikely	6.92% (5.82–8.15)	6.40% (5.15–7.83)	9.86% (8.24–11.68)
Unlikely	1.41% (1.01–1.92)	1.01% (0.59–1.61)	3.66% (2.71–4.82)
Very Unlikely	2.01% (1.49–2.63)	1.84% (1.26–2.59)	2.94% (2.02–4.11)

NOTE: Includes estimates for active-component Coast Guard members. 95-percent confidence intervals for each estimate are included in parentheses.

Table C.10.b
Perceived likelihood of respondent taking specific actions related to sexual assault or harassment, by service

Longform10a–g: How likely would you be to...

	Total DoD	Army	Navy	Air Force	Marine Corps	Coast Guard
Encourage someone who has experienced sexual harassment to tell a supervisor?						
Very likely	71.34% (69.85–72.80)	73.15% (70.82–75.39)	69.17% (65.14–73.00)	72.01% (70.55–73.43)	68.96% (64.15–73.49)	79.94% (78.15–81.64)
Likely	19.06% (17.83–20.33)	18.00% (16.03–20.12)	20.54% (17.64–23.69)	18.46% (17.21–19.75)	20.43% (16.36–25.01)	14.48% (12.98–16.08)
Neither likely nor unlikely	6.89% (5.88–8.02)	5.96% (4.73–7.41)	7.99% (4.90–12.15)	6.48% (5.74–7.29)	8.25% (5.81–11.30)	3.75% (2.98–4.64)
Unlikely	1.10% (0.88–1.35)	0.97% (0.67–1.35)	1.20% (0.67–1.97)	1.31% (1.00–1.69)	0.92% (0.40–1.81)	0.65% (0.37–1.05)
Very unlikely	1.61% (1.33–1.94)	1.91% (1.34–2.65)	1.10% (0.73–1.60)	1.74% (1.37–2.18)	1.43% (0.78–2.41)	1.19% (0.81–1.69)
Encourage someone who has experienced sexual assault to seek counseling?						
Very likely	77.55% (76.07–78.98)	77.15% (74.78–79.39)	76.74% (72.65–80.49)	81.09% (79.77–82.36)	74.08% (69.48–78.32)	83.49% (81.69–85.18)
Likely	16.30% (14.97–17.70)	14.87% (12.98–16.93)	19.21% (15.47–23.41)	14.30% (13.19–15.47)	18.61% (14.80–22.91)	12.54% (11.02–14.18)
Neither likely nor unlikely	4.59% (3.93–5.33)	6.00% (4.65–7.61)	3.11% (2.28–4.13)	3.22% (2.62–3.90)	5.57% (3.65–8.10)	2.78% (2.06–3.67)
Unlikely	0.34% (0.18–0.59)	0.39% (0.20–0.68)	0.10% (0.02–0.29)	0.22% (0.09–0.44)	0.81% (0.09–2.96)	0.24% (0.08–0.58)
Very unlikely	1.22% (0.96–1.52)	1.59% (1.04–2.31)	0.83% (0.51–1.29)	1.17% (0.83–1.60)	0.93% (0.49–1.59)	0.95% (0.59–1.43)
Encourage someone who has experienced sexual assault to report it?						
Very likely	77.69% (76.28–79.06)	77.48% (75.15–79.68)	77.15% (73.41–80.59)	79.80% (78.48–81.08)	75.68% (71.11–79.85)	84.74% (83.10–86.29)
Likely	15.78% (14.51–17.12)	15.25% (13.31–17.36)	17.22% (13.85–21.03)	14.70% (13.58–15.89)	16.61% (12.84–20.97)	11.01% (9.67–12.47)
Neither likely nor unlikely	4.84% (4.26–5.49)	5.15% (3.98–6.54)	4.70% (3.68–5.91)	4.00% (3.41–4.65)	5.67% (4.14–7.55)	2.85% (2.16–3.70)
Unlikely	0.42% (0.21–0.75)	0.44% (0.24–0.73)	0.11% (0.02–0.29)	0.30% (0.16–0.51)	1.10% (0.11–4.14)	0.29% (0.11–0.64)
Very unlikely	1.26% (1.00–1.56)	1.68% (1.13–2.40)	0.83% (0.51–1.28)	1.20% (0.86–1.62)	0.94% (0.51–1.60)	1.10% (0.73–1.60)
Tell a supervisor about sexual harassment if it happened to you?						
Very likely	61.11% (59.61–62.59)	62.72% (60.31–65.08)	59.42% (55.53–63.23)	60.37% (58.79–61.93)	60.83% (56.13–65.39)	66.63% (64.54–68.66)
Likely	21.06% (19.74–22.43)	20.15% (18.12–22.31)	22.58% (19.01–26.47)	21.06% (19.74–22.42)	20.94% (17.28–24.99)	19.76% (18.06–21.55)
Neither likely nor unlikely	10.46% (9.57–11.40)	9.80% (8.46–11.28)	11.69% (9.29–14.44)	10.20% (9.30–11.16)	10.62% (8.21–13.45)	8.10% (6.97–9.34)
Unlikely	3.72% (3.22–4.26)	3.51% (2.62–4.60)	3.33% (2.60–4.19)	4.35% (3.78–4.97)	3.86% (2.14–6.37)	3.33% (2.61–4.18)
Very unlikely	3.66% (3.15–4.22)	3.82% (2.87–4.97)	2.98% (2.18–3.98)	4.03% (3.44–4.68)	3.74% (2.22–5.86)	2.18% (1.65–2.83)

Table C.10.b—Continued

	Total DoD	Army	Navy	Air Force	Marine Corps	Coast Guard
Report a sexual assault if it happened to you?						
Very likely	67.47% (65.98–68.94)	67.77% (65.34–70.13)	66.94% (63.02–70.69)	68.62% (67.14–70.07)	65.65% (60.91–70.16)	73.22% (71.24–75.14)
Likely	19.34% (18.05–20.67)	18.09% (16.13–20.18)	20.86% (17.44–24.63)	19.01% (17.79–20.27)	20.68% (16.83–24.97)	16.44% (14.85–18.14)
Neither likely nor unlikely	7.90% (7.14–8.72)	8.14% (6.77–9.69)	7.22% (5.81–8.85)	7.74% (6.96–8.59)	8.68% (6.20–11.76)	6.92% (5.82–8.15)
Unlikely	2.47% (1.89–3.17)	2.75% (1.88–3.87)	2.84% (1.18–5.66)	2.12% (1.74–2.56)	1.68% (1.04–2.56)	1.41% (1.01–1.92)
Very unlikely	2.81% (2.35–3.34)	3.25% (2.36–4.36)	2.13% (1.52–2.91)	2.50% (2.02–3.06)	3.31% (1.83–5.45)	2.01% (1.49–2.63)

NOTE: Includes estimates for active-component DoD and Coast Guard members. 95-percent confidence intervals for each estimate are included in parentheses.

Table C.10.c

Perceived likelihood of respondent taking specific actions related to sexual assault or harassment, by pay grade

Longform10a–g: How likely would you be to...

	Total	E1–E4	E5–E9	O1–O3	O4–O6
Encourage someone who has experienced sexual harassment to tell a supervisor?					
Very likely	79.94% (78.15–81.64)	74.71% (70.86–78.28)	84.45% (82.25–86.48)	67.07% (60.64–73.05)	87.25% (82.28–91.24)
Likely	14.48% (12.98–16.08)	18.38% (15.29–21.80)	10.63% (8.90–12.57)	26.82% (20.98–33.32)	8.77% (5.50–13.13)
Neither likely nor unlikely	3.75% (2.98–4.64)	5.51% (3.69–7.87)	2.98% (2.13–4.04)	3.74% (2.14–6.04)	1.82% (0.60–4.18)
Unlikely	0.65% (0.37–1.05)	0.92% (0.30–2.11)	0.47% (0.19–0.98)	1.10% (0.34–2.62)	0.12% (0.00–1.55)
Very unlikely	1.19% (0.81–1.69)	0.49% (0.10–1.44)	1.47% (0.88–2.31)	1.27% (0.44–2.86)	2.04% (0.61–4.93)
Encourage someone who has experienced sexual assault to seek counseling?					
Very likely	83.49% (81.69–85.18)	76.23% (72.34–79.83)	86.55% (84.20–88.66)	83.84% (77.63–88.89)	91.83% (87.16–95.21)
Likely	12.54% (11.02–14.18)	17.65% (14.52–21.14)	10.17% (8.25–12.36)	14.20% (9.32–20.39)	5.58% (2.83–9.73)
Neither likely nor unlikely	2.78% (2.06–3.67)	5.41% (3.53–7.89)	1.79% (1.11–2.72)	1.41% (0.39–3.56)	0.67% (0.05–2.82)
Unlikely	0.24% (0.08–0.58)	0.26% (0.01–1.41)	0.30% (0.08–0.77)	0.09% (0.00–1.05)	0.00% (0.00–1.68)
Very unlikely	0.95% (0.59–1.43)	0.45% (0.08–1.43)	1.19% (0.66–1.97)	0.45% (0.03–1.98)	1.93% (0.53–4.86)
Encourage someone who has experienced sexual assault to report it?					
Very likely	84.74% (83.10–86.29)	78.74% (75.02–82.13)	88.38% (86.41–90.16)	79.17% (73.16–84.36)	91.62% (87.34–94.82)
Likely	11.01% (9.67–12.47)	15.74% (12.82–19.04)	7.90% (6.42–9.61)	16.80% (11.84–22.81)	5.40% (2.92–9.02)
Neither likely nor unlikely	2.85% (2.16–3.70)	4.73% (3.00–7.05)	2.08% (1.36–3.04)	2.50% (1.23–4.48)	0.91% (0.15–2.87)
Unlikely	0.29% (0.11–0.64)	0.26% (0.01–1.41)	0.29% (0.07–0.79)	0.53% (0.06–1.91)	0.14% (0.00–1.59)
Very unlikely	1.10% (0.73–1.60)	0.53% (0.12–1.46)	1.34% (0.77–2.16)	1.00% (0.29–2.49)	1.93% (0.53–4.86)
Tell a supervisor about sexual harassment if it happened to you?					
Very likely	66.63% (64.54–68.66)	62.54% (58.31–66.63)	72.25% (69.51–74.87)	49.69% (43.39–56.00)	66.63% (59.70–73.07)
Likely	19.76% (18.06–21.55)	21.75% (18.38–25.42)	16.41% (14.26–18.73)	31.43% (25.49–37.85)	19.53% (14.41–25.53)
Neither likely nor unlikely	8.10% (6.97–9.34)	11.55% (9.00–14.54)	5.55% (4.34–6.96)	10.55% (7.03–15.05)	8.44% (5.17–12.84)
Unlikely	3.33% (2.61–4.18)	2.88% (1.76–4.44)	3.22% (2.16–4.59)	5.28% (3.25–8.04)	3.35% (1.57–6.21)
Very unlikely	2.18% (1.65–2.83)	1.28% (0.60–2.37)	2.58% (1.77–3.61)	3.05% (1.55–5.35)	2.04% (0.61–4.93)

Table C.10.c—Continued

	Total	E1–E4	E5–E9	O1–O3	O4–O6
Report a sexual assault if it happened to you?					
Very likely	73.22% (71.24–75.14)	68.79% (64.72–72.67)	77.47% (74.87–79.91)	62.91% (56.45–69.05)	75.57% (69.23–81.19)
Likely	16.44% (14.85–18.14)	18.29% (15.15–21.77)	14.18% (12.15–16.40)	23.65% (18.18–29.86)	15.08% (10.61–20.53)
Neither likely nor unlikely	6.92% (5.82–8.15)	9.98% (7.57–12.85)	4.95% (3.70–6.46)	8.91% (5.47–13.53)	5.36% (2.83–9.10)
Unlikely	1.41% (1.01–1.92)	1.61% (0.83–2.79)	1.01% (0.56–1.66)	2.65% (1.21–4.98)	1.79% (0.58–4.14)
Very unlikely	2.01% (1.49–2.63)	1.33% (0.60–2.52)	2.40% (1.63–3.41)	1.87% (0.81–3.67)	2.19% (0.71–5.06)

NOTE: Includes estimates for active-component Coast Guard members. 95-percent confidence intervals for each estimate are included in parentheses.

Too few warrant officers were included in the sample to break them out as a separate pay grade. For the purposes of this table, warrant officers have been included with the E5–E9 category.

C.11. Percentage who observed a situation they believed was, or could have led to, a sexual assault

Table C.11.a
Percentage who observed a situation they believed was, or could have led to, a sexual assault, by gender

Longform11: In the past 12 months, did you observe a situation that you believed was, or could have led to, a sexual assault?

Total	Men	Women
3.35%	2.51%	8.11%
(2.69–4.13)	(1.80–3.40)	(6.62–9.80)

NOTE: Includes estimates for active-component Coast Guard members. 95-percent confidence intervals for each estimate are included in parentheses.

Table C.11.b
Percentage who observed a situation they believed was, or could have led to, a sexual assault, by service

Longform11: In the past 12 months, did you observe a situation that you believed was, or could have led to, a sexual assault?

Total DoD	Army	Navy	Air Force	Marine Corps	Coast Guard
6.77%	6.61%	9.09%	4.66%	6.82%	3.35%
(6.01–7.59)	(5.54–7.81)	(7.12–11.40)	(4.03–5.35)	(4.30–10.19)	(2.69–4.13)

NOTE: Includes estimates for active-component DoD and Coast Guard members. 95-percent confidence intervals for each estimate are included in parentheses.

Table C.11.c
Percentage who observed a situation they believed was, or could have led to, a sexual assault, by pay grade

Longform11: In the past 12 months, did you observe a situation that you believed was, or could have led to, a sexual assault?

Total	E1–E4	E5–E9	O1–O3	O4–O6
3.35%	4.93%	2.42%	3.99%	2.54%
(2.69–4.13)	(3.39–6.89)	(1.69–3.35)	(2.20–6.59)	(0.99–5.27)

NOTE: Includes estimates for active-component Coast Guard members. 95-percent confidence intervals for each estimate are included in parentheses.

Too few warrant officers were included in the sample to break them out as a separate pay grade. For the purposes of this table, warrant officers have been included with the E5–E9 category.

C.12. Type of bystander intervention behaviors taken among members who observed a situation they believed was, our could have led to, a sexual assault

Table C.12.a
Type of bystander intervention behaviors taken among members who observed a situation they believed was, our could have led to, a sexual assault, by gender

Longform11_1: Select the one response that most closely resembles your actions.

	Total	Men	Women
You stepped in and separated the people involved in the situation	18.76% (11.97–27.27)	NR (NR)	24.55% (16.49–34.17)
You asked the person who appeared to be at risk if they needed help	26.00% (16.83–37.02)	NR (NR)	21.72% (13.61–31.83)
You confronted the person who appeared to be causing the situation	19.23% (11.69–28.88)	NR (NR)	14.46% (8.01–23.31)
You created a distraction to cause one or more of the people to disengage from the situation	12.40% (6.16–21.51)	NR (NR)	14.89% (8.50–23.51)
You asked others to step in as a group and diffuse the situation	6.90% (2.65–14.19)	NR (NR)	9.23% (4.40–16.60)
You told someone in a position of authority about the situation	13.01% (6.41–22.66)	NR (NR)	10.46% (5.27–18.08)
You considered intervening in the situation, but you could not safely take any action	3.33% (0.57–10.22)	NR (NR)	3.69% (0.96–9.42)
You decided to not take action	0.36% (0.00–3.21)	NR (NR)	1.00% (0.03–5.42)

NOTE: Includes estimates for active-component Coast Guard members. 95-percent confidence intervals for each estimate are included in parentheses.

NR = Not reportable.

Table C.12.b
Type of bystander intervention behaviors taken among members who observed a situation they believed was, our could have led to, a sexual assault, by service

Longform11_1: Select the one response that most closely resembles your actions.

	Total DoD	Army	Navy	Air Force	Marine Corps	Coast Guard
You stepped in and separated the people involved in the situation	24.37% (19.22–30.13)	25.69% (17.31–35.63)	23.49% (13.51–36.21)	21.79% (16.34–28.07)	NR (12.64–43.19)	18.76% (11.97–27.27)
You asked the person who appeared to be at risk if they needed help	19.15% (14.84–24.08)	15.84% (10.83–22.00)	26.37% (16.01–39.07)	20.74% (15.47–26.85)	9.75% (3.97–19.17)	26.00% (16.83–37.02)
You confronted the person who appeared to be causing the situation	19.81% (14.04–26.68)	17.54% (11.76–24.68)	18.02% (8.66–31.31)	19.43% (13.29–26.88)	NR (7.89–63.09)	19.23% (11.69–28.88)
You created a distraction to cause one or more of the people to disengage from the situation	12.46% (9.65–15.73)	12.95% (8.13–19.23)	11.92% (6.98–18.60)	17.88% (12.93–23.77)	6.15% (2.26–13.02)	12.40% (6.16–21.51)
You asked others to step in as a group and diffuse the situation	3.75% (2.46–5.46)	3.71% (1.71–6.91)	3.71% (1.50–7.51)	4.44% (2.11–8.13)	3.18% (0.70–8.78)	6.90% (2.65–14.19)
You told someone in a position of authority about the situation	7.26% (5.32–9.64)	9.41% (5.97–13.95)	4.39% (2.11–7.96)	6.42% (3.80–10.04)	9.05% (2.84–20.43)	13.01% (6.41–22.66)
You considered intervening in the situation, but you could not safely take any action	4.48% (2.65–7.05)	5.10% (2.10–10.16)	4.49% (1.39–10.47)	2.87% (0.85–6.89)	4.64% (1.01–12.73)	3.33% (0.57–10.22)
You decided to not take action	8.72% (5.55–12.89)	9.75% (4.35–18.20)	7.61% (2.72–16.22)	6.43% (3.20–11.34)	11.10% (3.19–25.75)	0.36% (0.00–3.21)

NOTE: Includes estimates for active-component DoD and Coast Guard members. 95-percent confidence intervals for each estimate are included in parentheses.

NR = Not reportable.

Table C.12.c
Type of bystander intervention behaviors taken among members who observed a situation they believed was, our could have led to, a sexual assault, by pay grade

Longform11_1: Select the one response that most closely resembles your actions.

	Total	E1–E4	E5–E9	O1–O3	O4–O6
You stepped in and separated the people involved in the situation	18.76% (11.97–27.27)	7.61% (2.32–17.56)	NR (14.81–43.87)	NR (12.10–62.54)	NR (NR)
You asked the person who appeared to be at risk if they needed help	26.00% (16.83–37.02)	NR (19.27–54.09)	NR (11.01–39.70)	NR (1.47–28.71)	NR (NR)
You confronted the person who appeared to be causing the situation	19.23% (11.69–28.88)	NR (3.64–28.26)	NR (11.21–39.83)	NR (2.38–39.28)	NR (NR)
You created a distraction to cause one or more of the people to disengage from the situation	12.40% (6.16–21.51)	NR (5.07–33.00)	5.98% (1.32–16.10)	NR (2.34–48.63)	NR (NR)
You asked others to step in as a group and diffuse the situation	6.90% (2.65–14.19)	NR (2.79–26.51)	2.69% (0.17–11.34)	NR (1.04–27.06)	NR (NR)
You told someone in a position of authority about the situation	13.01% (6.41–22.66)	NR (2.83–28.73)	NR (5.81–33.24)	NR (2.05–40.00)	NR (NR)
You considered intervening in the situation, but you could not safely take any action	3.33% (0.57–10.22)	NR (0.60–21.02)	0.92% (0.00–8.44)	NR (0.01–18.71)	NR (NR)
You decided to not take action	0.36% (0.00–3.21)	0.80% (0.00–7.57)	0.00% (0.00–7.88)	NR (0.00–21.08)	NR (NR)

NOTE: Includes estimates for active-component Coast Guard members. 95-percent confidence intervals for each estimate are included in parentheses.

Too few warrant officers were included in the sample to break them out as a separate pay grade. For the purposes of this table, warrant officers have been included with the E5–E9 category.

NR = Not reportable.

C.13. Perceptions of unit leadership regarding sexual assault and harassment

Table C.13.a
Perceptions of unit leadership regarding sexual assault and harassment, by gender

Longform12a–e: Please indicate how well your unit leadership…

	Total	Men	Women
Makes it clear that sexual assault has no place in the military.			
Very well	81.37% (79.59–83.07)	83.10% (81.02–85.03)	71.65% (69.00–74.19)
Well	13.57% (12.10–15.16)	12.56% (10.88–14.40)	19.29% (17.07–21.67)
Neither well nor poorly	4.56% (3.65–5.61)	4.04% (3.02–5.29)	7.48% (6.07–9.10)
Poorly	0.27% (0.12–0.53)	0.12% (0.01–0.43)	1.15% (0.62–1.94)
Very poorly	0.22% (0.08–0.46)	0.18% (0.04–0.52)	0.43% (0.15–0.96)
Promotes a unit climate based on mutual respect and trust.			
Very well	72.91% (70.91–74.84)	75.51% (73.20–77.71)	58.30% (55.46–61.11)
Well	17.69% (16.06–19.42)	16.08% (14.22–18.08)	26.78% (24.28–29.40)
Neither well nor poorly	6.82% (5.75–8.03)	6.48% (5.25–7.89)	8.76% (7.23–10.49)
Poorly	1.68% (1.17–2.33)	1.17% (0.63–1.97)	4.56% (3.46–5.89)
Very poorly	0.89% (0.54–1.39)	0.77% (0.38–1.37)	1.60% (0.91–2.59)
Leads by example (e.g., refrains from sexist comments and behaviors).			
Very well	71.91% (69.89–73.88)	74.54% (72.18–76.78)	57.14% (54.29–59.96)
Well	18.24% (16.60–19.96)	16.69% (14.83–18.69)	26.93% (24.44–29.54)
Neither well nor poorly	7.29% (6.18–8.53)	6.85% (5.59–8.30)	9.76% (8.13–11.60)
Poorly	1.58% (0.95–2.48)	1.15% (0.48–2.32)	4.00% (2.96–5.27)
Very poorly	0.97% (0.62–1.45)	0.76% (0.39–1.33)	2.16% (1.35–3.27)

Table C.13.a—Continued

	Total	Men	Women
Catches and immediately corrects incidents of sexual harassment (e.g., inappropriate jokes, comments and behaviors).			
Very well	62.14% (59.97–64.28)	65.01% (62.49–67.47)	46.01% (43.18–48.86)
Well	23.86% (22.01–25.77)	22.72% (20.60–24.94)	30.27% (27.71–32.93)
Neither well nor poorly	11.48% (10.13–12.95)	10.43% (8.89–12.13)	17.42% (15.30–19.71)
Poorly	1.79% (1.13–2.69)	1.26% (0.57–2.41)	4.77% (3.58–6.22)
Very poorly	0.73% (0.43–1.15)	0.58% (0.27–1.11)	1.53% (0.92–2.36)
Creates an environment where victims would feel comfortable reporting sexual harassment or assault.			
Very well	68.69% (66.60–70.72)	71.56% (69.14–73.90)	52.45% (49.59–55.30)
Well	21.59% (19.82–23.43)	20.35% (18.33–22.48)	28.58% (26.04–31.22)
Neither well nor poorly	7.65% (6.42–9.03)	6.77% (5.37–8.39)	12.62% (10.79–14.65)
Poorly	1.32% (0.94–1.80)	0.83% (0.45–1.40)	4.11% (3.04–5.42)
Very poorly	0.76% (0.46–1.17)	0.50% (0.20–1.01)	2.23% (1.45–3.28)

NOTE: Includes estimates for active-component Coast Guard members. 95-percent confidence intervals for each estimate are included in parentheses.

Table C.13.b
Perceptions of unit leadership regarding sexual assault and harassment, by service

Longform12a–e: Please indicate how well your unit leadership…

	Total DoD	Army	Navy	Air Force	Marine Corps	Coast Guard
Makes it clear that sexual assault has no place in the military.						
Very well	78.87%	76.51%	77.04%	83.83%	80.12%	81.37%
	(77.67–80.04)	(74.43–78.49)	(74.05–79.84)	(82.56–85.04)	(76.26–83.60)	(79.59–83.07)
Well	14.12%	15.22%	15.84%	12.10%	11.61%	13.57%
	(13.15–15.14)	(13.55–17.01)	(13.38–18.56)	(11.09–13.16)	(9.23–14.35)	(12.10–15.16)
Neither well nor poorly	5.53%	6.81%	5.32%	3.07%	6.51%	4.56%
	(4.91–6.20)	(5.67–8.08)	(4.20–6.63)	(2.59–3.60)	(4.37–9.27)	(3.65–5.61)
Poorly	1.02%	1.00%	1.17%	0.74%	1.28%	0.27%
	(0.68–1.45)	(0.54–1.67)	(0.62–2.00)	(0.26–1.62)	(0.20–4.12)	(0.12–0.53)
Very poorly	0.46%	0.47%	0.63%	0.27%	0.48%	0.22%
	(0.28–0.72)	(0.23–0.86)	(0.17–1.58)	(0.12–0.51)	(0.07–1.62)	(0.08–0.46)
Promotes a unit climate based on mutual respect and trust.						
Very well	66.99%	65.65%	65.27%	72.99%	63.49%	72.91%
	(65.58–68.38)	(63.31–67.93)	(61.94–68.48)	(71.49–74.46)	(58.53–68.25)	(70.91–74.84)
Well	18.23%	17.97%	19.64%	17.12%	18.41%	17.69%
	(17.20–19.30)	(16.28–19.77)	(17.27–22.18)	(15.94–18.36)	(15.07–22.15)	(16.06–19.42)
Neither well nor poorly	8.91%	10.59%	8.14%	6.18%	10.25%	6.82%
	(7.98–9.91)	(8.87–12.52)	(6.75–9.70)	(5.40–7.03)	(6.86–14.55)	(5.75–8.03)
Poorly	2.80%	2.84%	3.05%	2.15%	3.36%	1.68%
	(2.39–3.26)	(2.26–3.51)	(2.21–4.08)	(1.73–2.64)	(1.76–5.78)	(1.17–2.33)
Very poorly	3.06%	2.95%	3.91%	1.55%	4.49%	0.89%
	(2.47–3.76)	(2.19–3.87)	(2.40–5.99)	(0.96–2.37)	(2.42–7.53)	(0.54–1.39)
Leads by example (e.g., refrains from sexist comments and behaviors).						
Very well	64.37%	62.01%	62.02%	74.39%	57.92%	71.91%
	(62.92–65.79)	(59.62–64.35)	(58.61–65.34)	(72.93–75.80)	(52.91–62.81)	(69.89–73.88)
Well	19.58%	20.91%	20.33%	16.60%	19.70%	18.24%
	(18.49–20.71)	(19.03–22.88)	(17.98–22.85)	(15.46–17.78)	(16.07–23.76)	(16.60–19.96)
Neither well nor poorly	9.92%	11.37%	9.35%	5.90%	13.69%	7.29%
	(8.97–10.94)	(9.66–13.28)	(7.89–10.98)	(5.17–6.71)	(10.01–18.11)	(6.18–8.53)
Poorly	3.42%	2.99%	4.96%	1.58%	5.08%	1.58%
	(2.88–4.03)	(2.37–3.73)	(3.40–6.95)	(1.25–1.97)	(3.24–7.54)	(0.95–2.48)
Very poorly	2.71%	2.72%	3.34%	1.53%	3.60%	0.97%
	(2.18–3.33)	(1.97–3.65)	(2.31–4.66)	(0.94–2.35)	(1.53–7.08)	(0.62–1.45)
Catches and immediately corrects incidents of sexual harassment (e.g., inappropriate jokes, comments, and behaviors).						
Very well	58.14%	56.25%	55.15%	66.18%	54.85%	62.14%
	(56.66–59.61)	(53.84–58.64)	(51.48–58.78)	(64.63–67.70)	(49.93–59.70)	(59.97–64.28)
Well	23.01%	24.42%	25.35%	19.82%	20.57%	23.86%
	(21.85–24.21)	(22.38–26.55)	(22.54–28.31)	(18.60–21.09)	(17.37–24.06)	(22.01–25.77)
Neither well nor poorly	13.86%	14.47%	12.93%	10.55%	19.29%	11.48%
	(12.76–15.01)	(12.61–16.50)	(11.17–14.86)	(9.53–11.64)	(14.89–24.34)	(10.13–12.95)
Poorly	3.14%	3.06%	4.13%	2.38%	2.93%	1.79%
	(2.64–3.70)	(2.46–3.77)	(2.72–6.00)	(1.89–2.96)	(1.54–5.02)	(1.13–2.69)
Very poorly	1.85%	1.79%	2.44%	1.07%	2.36%	0.73%
	(1.45–2.33)	(1.21–2.54)	(1.56–3.63)	(0.77–1.43)	(0.91–4.95)	(0.43–1.15)

Table C.13.b—Continued

	Total DoD	Army	Navy	Air Force	Marine Corps	Coast Guard
Creates an environment where victims would feel comfortable reporting sexual harassment or assault.						
Very well	64.86% (63.41–66.29)	63.39% (61.01–65.73)	62.85% (59.39–66.21)	71.43% (69.94–72.89)	61.22% (56.16–66.11)	68.69% (66.60–70.72)
Well	21.00% (19.82–22.21)	20.81% (18.88–22.84)	22.92% (20.22–25.80)	18.51% (17.31–19.77)	22.43% (18.40–26.87)	21.59% (19.82–23.43)
Neither well nor poorly	10.18% (9.21–11.21)	11.26% (9.61–13.07)	10.45% (8.54–12.63)	7.33% (6.41–8.34)	11.54% (8.12–15.76)	7.65% (6.42–9.03)
Poorly	2.31% (1.89–2.79)	2.81% (2.12–3.66)	1.84% (1.32–2.50)	1.60% (1.27–1.98)	2.90% (1.22–5.76)	1.32% (0.94–1.80)
Very poorly	1.66% (1.31–2.07)	1.73% (1.20–2.40)	1.93% (1.16–3.01)	1.13% (0.84–1.49)	1.91% (0.77–3.90)	0.76% (0.46–1.17)

NOTE: Includes estimates for active-component DoD and Coast Guard members. 95-percent confidence intervals for each estimate are included in parentheses.

Table C.13.c
Perceptions of unit leadership regarding sexual assault and harassment, by pay grade

Longform12a–e: Please indicate how well your unit leadership…

	Total	E1–E4	E5–E9	O1–O3	O4–O6
Makes it clear that sexual assault has no place in the military.					
Very well	81.37% (79.59–83.07)	77.29% (73.56–80.72)	83.54% (81.08–85.80)	77.89% (72.55–82.63)	87.89% (82.86–91.87)
Well	13.57% (12.10–15.16)	15.98% (13.07–19.24)	11.53% (9.58–13.71)	19.14% (14.60–24.38)	10.51% (6.79–15.33)
Neither well nor poorly	4.56% (3.65–5.61)	6.50% (4.52–9.01)	4.21% (3.04–5.68)	2.53% (1.25–4.52)	1.60% (0.44–4.04)
Poorly	0.27% (0.12–0.53)	0.19% (0.02–0.76)	0.37% (0.12–0.83)	0.25% (0.01–1.34)	0.00% (0.00–1.67)
Very poorly	0.22% (0.08–0.46)	0.04% (0.00–0.50)	0.36% (0.12–0.82)	0.19% (0.00–1.24)	0.00% (0.00–1.67)
Promotes a unit climate based on mutual respect and trust.					
Very well	72.91% (70.91–74.84)	66.73% (62.59–70.68)	76.02% (73.32–78.58)	69.36% (63.38–74.89)	81.50% (75.81–86.33)
Well	17.69% (16.06–19.42)	20.63% (17.42–24.15)	15.61% (13.44–17.99)	23.10% (18.03–28.82)	12.84% (8.77–17.93)
Neither well nor poorly	6.82% (5.75–8.03)	9.79% (7.30–12.79)	5.80% (4.53–7.28)	5.23% (3.06–8.26)	3.86% (1.87–6.98)
Poorly	1.68% (1.17–2.33)	1.26% (0.63–2.24)	1.84% (1.03–3.02)	2.15% (0.92–4.21)	1.68% (0.50–4.05)
Very poorly	0.89% (0.54–1.39)	1.59% (0.70–3.08)	0.73% (0.36–1.32)	0.17% (0.00–1.19)	0.12% (0.00–1.55)
Leads by example (e.g., refrains from sexist comments and behaviors).					
Very well	71.91% (69.89–73.88)	66.10% (61.97–70.05)	73.89% (71.09–76.55)	70.60% (64.69–76.04)	84.31% (78.61–89.00)
Well	18.24% (16.60–19.96)	20.40% (17.21–23.88)	16.95% (14.75–19.33)	22.11% (17.18–27.70)	13.00% (8.66–18.47)
Neither well nor poorly	7.29% (6.18–8.53)	10.50% (7.95–13.52)	6.63% (5.27–8.20)	4.63% (2.64–7.46)	2.09% (0.66–4.89)
Poorly	1.58% (0.95–2.48)	2.15% (1.26–3.42)	1.30% (0.39–3.15)	2.10% (0.70–4.77)	0.48% (0.03–2.20)
Very poorly	0.97% (0.62–1.45)	0.86% (0.29–1.94)	1.24% (0.70–2.01)	0.56% (0.09–1.84)	0.12% (0.00–1.56)
Catches and immediately corrects incidents of sexual harassment (e.g., inappropriate jokes, comments, and behaviors).					
Very well	62.14% (59.97–64.28)	58.07% (53.76–62.28)	65.68% (62.74–68.53)	51.99% (45.64–58.30)	68.00% (60.15–75.16)
Well	23.86% (22.01–25.77)	25.80% (22.23–29.63)	20.78% (18.44–23.26)	33.35% (27.25–39.89)	24.71% (17.86–32.65)
Neither well nor poorly	11.48% (10.13–12.95)	13.28% (10.52–16.44)	11.06% (9.25–13.09)	12.11% (8.66–16.34)	6.00% (3.29–9.92)
Poorly	1.79% (1.13–2.69)	1.91% (1.11–3.06)	1.81% (0.77–3.57)	2.13% (0.84–4.39)	0.69% (0.08–2.54)
Very poorly	0.73% (0.43–1.15)	0.94% (0.35–2.04)	0.68% (0.30–1.30)	0.41% (0.04–1.60)	0.59% (0.04–2.53)

Table C.13.c—Continued

	Total	E1–E4	E5–E9	O1–O3	O4–O6
Creates an environment where victims would feel comfortable reporting sexual harassment or assault.					
Very well	68.69% (66.60–70.72)	65.70% (61.56–69.66)	72.13% (69.33–74.80)	59.19% (52.74–65.42)	69.61% (61.64–76.80)
Well	21.59% (19.82–23.43)	22.54% (19.17–26.19)	18.53% (16.35–20.87)	31.36% (25.35–37.87)	26.08% (19.01–34.19)
Neither well nor poorly	7.65% (6.42–9.03)	9.17% (6.84–11.96)	7.55% (5.79–9.64)	6.40% (3.96–9.72)	3.66% (1.79–6.59)
Poorly	1.32% (0.94–1.80)	1.73% (0.98–2.84)	0.94% (0.51–1.59)	2.64% (1.13–5.19)	0.52% (0.03–2.25)
Very poorly	0.76% (0.46–1.17)	0.86% (0.30–1.94)	0.85% (0.44–1.47)	0.40% (0.04–1.59)	0.14% (0.00–1.58)

NOTE: Includes estimates for active-component Coast Guard members. 95-percent confidence intervals for each estimate are included in parentheses.

Too few warrant officers were included in the sample to break them out as a separate pay grade. For the purposes of this table, warrant officers have been included with the E5–E9 category.

C.14. Exposure to sexual assault prevention and response training

Table C.14.a
Exposure to sexual assault prevention and response training, by gender

Longform22: Have you had any military training during the past 12 months on topics related to sexual assault?

Total	Men	Women
98.78%	98.83%	98.54%
(98.21–99.21)	(98.15–99.31)	(97.71–99.13)

NOTE: Includes estimates for active-component Coast Guard members. 95-percent confidence intervals for each estimate are included in parentheses.

Table C.14.b
Exposure to sexual assault prevention and response training, by service

Longform22: Have you had any military training during the past 12 months on topics related to sexual assault?

Total DoD	Army	Navy	Air Force	Marine Corps	Coast Guard
97.89%	96.95%	98.80%	98.50%	97.85%	98.78%
(97.20–98.44)	(95.26–98.16)	(98.28–99.19)	(98.00–98.89)	(95.34–99.22)	(98.21–99.21)

NOTE: Includes estimates for active-component DoD and Coast Guard members. 95-percent confidence intervals for each estimate are included in parentheses.

Table C.14.c
Exposure to sexual assault prevention and response training, by pay grade

Longform22: Have you had any military training during the past 12 months on topics related to sexual assault?

Total	E1–E4	E5–E9	O1–O3	O4–O6
98.78%	98.60%	99.25%	97.91%	97.43%
(98.21–99.21)	(97.24–99.40)	(98.65–99.63)	(95.43–99.25)	(92.47–99.50)

NOTE: Includes estimates for active-component Coast Guard members. 95-percent confidence intervals for each estimate are included in parentheses.

Too few warrant officers were included in the sample to break them out as a separate pay grade. For the purposes of this table, warrant officers have been included with the E5–E9 category.

C.15. Perception of sexual assault prevention and response training

Table C.15.a
Perception of sexual assault prevention and response training, by gender

Longform23a–j: My Service's sexual assault training...

	Total	Men	Women
Provides a good understanding of what actions are considered sexual assault.			
Strongly disagree	1.49% (1.01–2.10)	1.48% (0.95–2.21)	1.50% (0.86–2.43)
Disagree	0.81% (0.49–1.24)	0.75% (0.39–1.29)	1.17% (0.59–2.07)
Neither agree nor disagree	5.86% (4.82–7.04)	5.82% (4.63–7.21)	6.03% (4.75–7.55)
Agree	44.20% (41.97–46.46)	43.73% (41.15–46.33)	46.91% (44.01–49.82)
Strongly agree	47.65% (45.39–49.91)	48.22% (45.62–50.83)	44.39% (41.51–47.30)
Teaches that consumption of alcohol may increase the likelihood of sexual assault.			
Strongly disagree	1.55% (1.07–2.19)	1.59% (1.03–2.33)	1.36% (0.76–2.23)
Disagree	0.35% (0.16–0.66)	0.30% (0.10–0.70)	0.64% (0.21–1.46)
Neither agree nor disagree	5.60% (4.59–6.77)	5.62% (4.45–6.98)	5.53% (4.28–7.02)
Agree	40.03% (37.84–42.25)	39.62% (37.10–42.19)	42.38% (39.52–45.27)
Strongly agree	52.46% (50.20–54.71)	52.87% (50.26–55.47)	50.10% (47.18–53.01)
Teaches how to avoid situations that might increase risk of being a victim of sexual assault.			
Strongly disagree	1.79% (1.28–2.45)	1.79% (1.20–2.55)	1.84% (1.14–2.81)
Disagree	1.16% (0.77–1.68)	1.13% (0.68–1.76)	1.36% (0.75–2.28)
Neither agree nor disagree	7.45% (6.28–8.75)	7.26% (5.93–8.78)	8.51% (6.97–10.25)
Agree	43.63% (41.41–45.88)	42.98% (40.41–45.58)	47.35% (44.44–50.26)
Strongly agree	45.97% (43.72–48.23)	46.85% (44.25–49.46)	40.94% (38.10–43.83)

Table C.15.a—Continued

	Total	Men	Women
Teaches how to intervene when you witness a situation involving a fellow Service member (bystander intervention).			
Strongly disagree	1.99% (1.44–2.67)	1.95% (1.33–2.76)	2.18% (1.41–3.22)
Disagree	1.44% (0.97–2.05)	1.38% (0.86–2.11)	1.73% (1.05–2.68)
Neither agree nor disagree	8.61% (7.39–9.95)	8.53% (7.14–10.09)	9.02% (7.42–10.83)
Agree	44.17% (41.93–46.42)	43.44% (40.85–46.05)	48.31% (45.41–51.23)
Strongly agree	43.81% (41.57–46.06)	44.69% (42.11–47.29)	38.76% (35.95–41.62)
Teaches how to obtain medical care following a sexual assault.			
Strongly disagree	1.73% (1.21–2.40)	1.72% (1.12–2.51)	1.83% (1.13–2.79)
Disagree	2.36% (1.79–3.06)	2.07% (1.43–2.89)	4.02% (2.95–5.34)
Neither agree nor disagree	9.34% (8.07–10.73)	9.00% (7.56–10.62)	11.24% (9.48–13.19)
Agree	43.76% (41.53–46.01)	43.46% (40.88–46.06)	45.47% (42.59–48.38)
Strongly agree	42.81% (40.58–45.06)	43.75% (41.17–46.36)	37.44% (34.64–40.31)
Explains the role of the chain of command in handling sexual assaults.			
Strongly disagree	1.75% (1.22–2.43)	1.73% (1.14–2.53)	1.86% (1.15–2.84)
Disagree	1.46% (1.03–2.00)	1.08% (0.63–1.71)	3.64% (2.61–4.92)
Neither agree nor disagree	7.81% (6.66–9.08)	7.39% (6.09–8.87)	10.18% (8.47–12.11)
Agree	44.83% (42.59–47.09)	44.48% (41.89–47.09)	46.85% (43.95–49.76)
Strongly agree	44.15% (41.90–46.41)	45.32% (42.72–47.94)	37.48% (34.68–40.34)
Explains the reporting options available if a sexual assault occurs.			
Strongly disagree	1.57% (1.07–2.22)	1.59% (1.02–2.35)	1.49% (0.85–2.41)
Disagree	0.77% (0.47–1.19)	0.72% (0.37–1.25)	1.06% (0.54–1.89)
Neither agree nor disagree	6.32% (5.27–7.52)	6.24% (5.03–7.64)	6.79% (5.40–8.41)
Agree	43.73% (41.50–45.98)	42.92% (40.35–45.52)	48.34% (45.43–51.26)
Strongly agree	47.60% (45.35–49.87)	48.53% (45.92–51.14)	42.31% (39.45–45.21)

Table C.15.a—Continued

	Total	Men	Women
Identifies the points of contact for reporting sexual assault (e.g., SARC, Victim Advocate).			
Strongly disagree	1.65% (1.14–2.32)	1.70% (1.11–2.48)	1.40% (0.79–2.31)
Disagree	0.93% (0.61–1.37)	0.74% (0.38–1.28)	2.05% (1.25–3.16)
Neither agree nor disagree	7.35% (6.21–8.61)	7.18% (5.89–8.65)	8.29% (6.77–10.03)
Agree	43.81% (41.58–46.06)	43.43% (40.86–46.03)	45.96% (43.07–48.87)
Strongly agree	46.26% (44.01–48.52)	46.95% (44.35–49.57)	42.30% (39.44–45.20)
Explains how sexual assault is a mission readiness problem.			
Strongly disagree	1.74% (1.22–2.40)	1.70% (1.11–2.48)	1.96% (1.22–2.97)
Disagree	1.06% (0.71–1.51)	0.74% (0.39–1.28)	2.86% (1.98–4.00)
Neither agree nor disagree	8.10% (6.88–9.46)	7.94% (6.55–9.53)	8.97% (7.38–10.79)
Agree	42.58% (40.36–44.82)	42.06% (39.51–44.65)	45.55% (42.66–48.46)
Strongly agree	46.53% (44.28–48.79)	47.56% (44.95–50.17)	40.66% (37.81–43.55)
Explains the resources available to victims (e.g., Safe Helpline).			
Strongly disagree	1.73% (1.21–2.40)	1.75% (1.15–2.53)	1.63% (0.96–2.57)
Disagree	1.16% (0.79–1.65)	0.86% (0.47–1.43)	2.89% (1.96–4.10)
Neither agree nor disagree	7.18% (5.97–8.55)	7.02% (5.63–8.62)	8.12% (6.63–9.82)
Agree	44.06% (41.84–46.30)	43.40% (40.84–46.00)	47.79% (44.89–50.71)
Strongly agree	45.87% (43.61–48.13)	46.97% (44.37–49.59)	39.57% (36.74–42.45)

NOTE: Includes estimates for active-component Coast Guard members. 95-percent confidence intervals for each estimate are included in parentheses.

Table C.15.b
Perception of sexual assault prevention and response training, by service

Longform23a–j: My Service's sexual assault training…

	Total DoD	Army	Navy	Air Force	Marine Corps	Coast Guard
Provides a good understanding of what actions are considered sexual assault.						
Strongly disagree	1.62% (1.33–1.96)	1.51% (1.07–2.07)	1.57% (0.91–2.52)	1.81% (1.41–2.28)	1.70% (0.86–3.00)	1.49% (1.01–2.10)
Disagree	0.79% (0.57–1.05)	0.72% (0.37–1.27)	1.06% (0.59–1.74)	0.67% (0.45–0.95)	0.70% (0.19–1.79)	0.81% (0.49–1.24)
Neither agree nor disagree	7.27% (6.53–8.07)	9.59% (8.13–11.21)	5.99% (4.78–7.40)	4.92% (4.07–5.89)	7.29% (5.12–10.01)	5.86% (4.82–7.04)
Agree	37.18% (35.68–38.70)	38.48% (36.16–40.84)	41.21% (37.22–45.29)	33.08% (31.58–34.60)	33.79% (29.36–38.44)	44.20% (41.97–46.46)
Strongly agree	53.14% (51.60–54.67)	49.70% (47.29–52.12)	50.16% (46.20–54.13)	59.53% (57.90–61.14)	56.53% (51.66–61.30)	47.65% (45.39–49.91)
Teaches that consumption of alcohol may increase the likelihood of sexual assault.						
Strongly disagree	1.52% (1.25–1.83)	1.49% (1.05–2.05)	1.20% (0.68–1.96)	1.92% (1.51–2.40)	1.46% (0.72–2.62)	1.55% (1.07–2.19)
Disagree	0.85% (0.46–1.43)	0.76% (0.37–1.38)	0.58% (0.20–1.31)	0.68% (0.47–0.95)	1.85% (0.17–7.16)	0.35% (0.16–0.66)
Neither agree nor disagree	7.11% (6.40–7.88)	9.49% (8.07–11.07)	5.91% (4.73–7.28)	4.64% (3.99–5.37)	7.04% (4.90–9.75)	5.60% (4.59–6.77)
Agree	35.05% (33.56–36.56)	36.35% (34.06–38.68)	38.24% (34.24–42.36)	31.78% (30.25–33.33)	31.69% (27.35–36.28)	40.03% (37.84–42.25)
Strongly agree	55.47% (53.93–57.00)	51.91% (49.49–54.32)	54.07% (50.06–58.05)	60.98% (59.37–62.58)	57.96% (53.00–62.80)	52.46% (50.20–54.71)
Teaches how to avoid situations that might increase risk of being a victim of sexual assault.						
Strongly disagree	1.80% (1.52–2.11)	1.53% (1.12–2.04)	1.54% (0.96–2.34)	2.52% (2.05–3.05)	1.72% (0.93–2.89)	1.79% (1.28–2.45)
Disagree	1.43% (1.14–1.77)	1.53% (1.03–2.17)	1.38% (0.74–2.34)	1.55% (1.22–1.95)	1.02% (0.37–2.22)	1.16% (0.77–1.68)
Neither agree nor disagree	8.61% (7.84–9.44)	11.32% (9.79–13.00)	7.46% (6.10–9.02)	5.99% (5.26–6.79)	7.84% (5.62–10.60)	7.45% (6.28–8.75)
Agree	36.49% (35.00–38.00)	36.46% (34.20–38.78)	39.26% (35.36–43.27)	33.81% (32.26–35.38)	36.39% (31.57–41.43)	43.63% (41.41–45.88)
Strongly agree	51.67% (50.14–53.20)	49.16% (46.74–51.58)	50.36% (46.40–54.32)	56.13% (54.49–57.77)	53.03% (48.04–57.97)	45.97% (43.72–48.23)
Teaches how to intervene when you witness a situation involving a fellow Service member (bystander intervention).						
Strongly disagree	1.50% (1.23–1.81)	1.33% (0.94–1.83)	1.40% (0.83–2.22)	1.88% (1.43–2.42)	1.49% (0.75–2.66)	1.99% (1.44–2.67)
Disagree	0.99% (0.74–1.29)	1.04% (0.62–1.64)	1.02% (0.54–1.77)	0.85% (0.61–1.16)	1.01% (0.35–2.25)	1.44% (0.97–2.05)
Neither agree nor disagree	8.12% (7.34–8.96)	10.58% (9.07–12.24)	6.55% (5.27–8.03)	5.74% (5.01–6.55)	8.33% (5.78–11.54)	8.61% (7.39–9.95)
Agree	37.05% (35.56–38.56)	37.76% (35.47–40.10)	40.32% (36.38–44.35)	34.01% (32.47–35.57)	34.71% (30.03–39.63)	44.17% (41.93–46.42)
Strongly agree	52.34% (50.81–53.87)	49.29% (46.87–51.70)	50.71% (46.74–54.66)	57.52% (55.89–59.15)	54.45% (49.47–59.36)	43.81% (41.57–46.06)

Table C.15.b—Continued

	Total DoD	Army	Navy	Air Force	Marine Corps	Coast Guard
Teaches how to obtain medical care following a sexual assault.						
Strongly disagree	1.71% (1.41–2.05)	1.52% (1.10–2.04)	1.90% (1.14–2.95)	1.89% (1.49–2.37)	1.57% (0.80–2.73)	1.73% (1.21–2.40)
Disagree	1.32% (1.07–1.61)	1.29% (0.86–1.85)	1.11% (0.68–1.71)	1.68% (1.31–2.14)	1.14% (0.45–2.37)	2.36% (1.79–3.06)
Neither agree nor disagree	9.18% (8.40–10.01)	11.86% (10.33–13.53)	7.55% (6.17–9.13)	6.92% (6.12–7.78)	8.70% (6.41–11.49)	9.34% (8.07–10.73)
Agree	36.83% (35.32–38.37)	37.63% (35.32–39.98)	40.60% (36.56–44.72)	33.62% (32.08–35.19)	33.71% (29.09–38.56)	43.76% (41.53–46.01)
Strongly agree	50.95% (49.42–52.49)	47.70% (45.29–50.12)	48.85% (44.90–52.80)	55.88% (54.24–57.52)	54.88% (49.93–59.76)	42.81% (40.58–45.06)
Explains the role of the chain of command in handling sexual assaults.						
Strongly disagree	1.62% (1.34–1.95)	1.44% (1.03–1.94)	1.66% (0.98–2.63)	1.83% (1.44–2.30)	1.68% (0.89–2.87)	1.75% (1.22–2.43)
Disagree	1.15% (0.90–1.45)	1.14% (0.74–1.69)	1.51% (0.92–2.34)	0.98% (0.72–1.31)	0.84% (0.28–1.90)	1.46% (1.03–2.00)
Neither agree nor disagree	8.56% (7.77–9.40)	11.32% (9.76–13.02)	7.06% (5.82–8.48)	5.83% (5.11–6.62)	8.45% (5.90–11.65)	7.81% (6.66–9.08)
Agree	36.94% (35.44–38.46)	37.44% (35.16–39.77)	40.41% (36.36–44.55)	34.51% (32.95–36.10)	33.77% (29.39–38.37)	44.83% (42.59–47.09)
Strongly agree	51.72% (50.19–53.26)	48.66% (46.25–51.08)	49.36% (45.40–53.32)	56.85% (55.20–58.48)	55.26% (50.38–60.06)	44.15% (41.90–46.41)
Explains the reporting options available if a sexual assault occurs.						
Strongly disagree	1.41% (1.15–1.70)	1.36% (0.96–1.86)	1.23% (0.70–1.99)	1.64% (1.26–2.09)	1.47% (0.73–2.63)	1.57% (1.07–2.22)
Disagree	0.60% (0.40–0.88)	0.68% (0.33–1.26)	0.58% (0.19–1.31)	0.41% (0.25–0.66)	0.75% (0.21–1.86)	0.77% (0.47–1.19)
Neither agree nor disagree	7.28% (6.54–8.07)	9.89% (8.41–11.52)	6.24% (5.00–7.68)	4.50% (3.84–5.24)	6.86% (4.71–9.60)	6.32% (5.27–7.52)
Agree	35.70% (34.22–37.21)	36.57% (34.30–38.88)	38.06% (34.18–42.06)	32.01% (30.50–33.56)	35.62% (30.84–40.63)	43.73% (41.50–45.98)
Strongly agree	55.01% (53.47–56.53)	51.50% (49.09–53.92)	53.89% (49.92–57.83)	61.43% (59.82–63.03)	55.30% (50.30–60.22)	47.60% (45.35–49.87)
Identifies the points of contact for reporting sexual assault (e.g., SARC, Victim Advocate).						
Strongly disagree	1.47% (1.19–1.78)	1.44% (1.03–1.97)	1.35% (0.73–2.27)	1.61% (1.24–2.06)	1.48% (0.74–2.63)	1.65% (1.14–2.32)
Disagree	0.57% (0.39–0.81)	0.73% (0.38–1.27)	0.59% (0.32–1.00)	0.39% (0.21–0.64)	0.42% (0.04–1.65)	0.93% (0.61–1.37)
Neither agree nor disagree	7.38% (6.58–8.24)	10.11% (8.65–11.72)	5.54% (4.41–6.86)	4.07% (3.45–4.78)	8.93% (5.81–12.99)	7.35% (6.21–8.61)
Agree	36.04% (34.55–37.55)	37.20% (34.88–39.57)	40.16% (36.22–44.21)	32.02% (30.50–33.57)	32.70% (28.28–37.37)	43.81% (41.58–46.06)
Strongly agree	54.54% (53.00–56.07)	50.52% (48.10–52.93)	52.36% (48.38–56.31)	61.91% (60.30–63.50)	56.47% (51.50–61.34)	46.26% (44.01–48.52)

Table C.15.b—Continued

	Total DoD	Army	Navy	Air Force	Marine Corps	Coast Guard
Explains how sexual assault is a mission readiness problem.						
Strongly disagree	1.58% (1.30–1.90)	1.49% (1.08–2.01)	1.59% (0.93–2.53)	1.79% (1.40–2.26)	1.44% (0.70–2.60)	1.74% (1.22–2.40)
Disagree	0.92% (0.68–1.22)	0.95% (0.51–1.59)	0.88% (0.45–1.56)	0.82% (0.59–1.11)	1.11% (0.44–2.29)	1.06% (0.71–1.51)
Neither agree nor disagree	8.01% (7.20–8.88)	10.30% (8.82–11.92)	5.86% (4.71–7.18)	5.84% (5.10–6.65)	9.37% (6.22–13.40)	8.10% (6.88–9.46)
Agree	35.93% (34.45–37.44)	36.83% (34.55–39.16)	40.28% (36.29–44.37)	32.65% (31.12–34.21)	31.63% (27.37–36.14)	42.58% (40.36–44.82)
Strongly agree	53.55% (52.01–55.08)	50.43% (48.01–52.84)	51.40% (47.42–55.36)	58.90% (57.27–60.52)	56.45% (51.52–61.29)	46.53% (44.28–48.79)
Explains the resources available to victims (e.g., Safe Helpline).						
Strongly disagree	1.59% (1.31–1.92)	1.54% (1.11–2.07)	1.56% (0.89–2.53)	1.75% (1.37–2.21)	1.49% (0.75–2.65)	1.73% (1.21–2.40)
Disagree	0.78% (0.57–1.05)	0.75% (0.40–1.28)	0.96% (0.53–1.61)	0.62% (0.42–0.88)	0.82% (0.22–2.08)	1.16% (0.79–1.65)
Neither agree nor disagree	7.85% (7.04–8.73)	10.61% (9.12–12.26)	5.85% (4.70–7.18)	5.04% (4.34–5.82)	8.72% (5.62–12.79)	7.18% (5.97–8.55)
Agree	36.39% (34.88–37.91)	37.01% (34.72–39.35)	41.10% (37.06–45.23)	33.34% (31.80–34.91)	31.80% (27.53–36.31)	44.06% (41.84–46.30)
Strongly agree	53.39% (51.85–54.93)	50.09% (47.67–52.51)	50.53% (46.55–54.50)	59.24% (57.61–60.86)	57.17% (52.24–61.99)	45.87% (43.61–48.13)

NOTE: Includes estimates for active-component DoD and Coast Guard members. 95-percent confidence intervals for each estimate are included in parentheses.

Table C.15.c
Perception of sexual assault prevention and response training, by pay grade

Longform23a–j: My Service's sexual assault training…

	Total	E1–E4	E5–E9	O1–O3	O4–O6
Provides a good understanding of what actions are considered sexual assault.					
Strongly disagree	1.49%	1.81%	1.39%	2.00%	0.17%
	(1.01–2.10)	(0.83–3.41)	(0.81–2.21)	(0.73–4.35)	(0.00–1.66)
Disagree	0.81%	0.49%	0.97%	0.27%	1.72%
	(0.49–1.24)	(0.10–1.44)	(0.51–1.65)	(0.01–1.42)	(0.40–4.66)
Neither agree nor disagree	5.86%	8.60%	4.98%	3.32%	4.26%
	(4.82–7.04)	(6.19–11.57)	(3.78–6.42)	(1.24–7.07)	(1.98–7.90)
Agree	44.20%	41.08%	45.93%	49.54%	37.57%
	(41.97–46.46)	(36.74–45.53)	(42.87–49.00)	(43.10–55.99)	(30.18–45.42)
Strongly agree	47.65%	48.02%	46.74%	44.86%	56.28%
	(45.39–49.91)	(43.51–52.56)	(43.70–49.81)	(38.46–51.39)	(48.57–63.78)
Teaches that consumption of alcohol may increase the likelihood of sexual assault.					
Strongly disagree	1.55%	1.78%	1.44%	2.10%	0.69%
	(1.07–2.19)	(0.80–3.39)	(0.85–2.29)	(0.80–4.43)	(0.06–2.77)
Disagree	0.35%	0.23%	0.52%	0.10%	0.00%
	(0.16–0.66)	(0.01–1.05)	(0.21–1.05)	(0.00–1.11)	(0.00–1.70)
Neither agree nor disagree	5.60%	8.09%	4.68%	4.40%	3.66%
	(4.59–6.77)	(5.75–10.99)	(3.51–6.10)	(2.35–7.44)	(1.14–8.56)
Agree	40.03%	38.75%	42.64%	40.36%	26.21%
	(37.84–42.25)	(34.45–43.17)	(39.63–45.69)	(34.15–46.81)	(20.33–32.79)
Strongly agree	52.46%	51.16%	50.72%	53.04%	69.45%
	(50.20–54.71)	(46.63–55.68)	(47.66–53.78)	(46.58–59.43)	(62.41–75.87)
Teaches how to avoid situations that might increase risk of being a victim of sexual assault.					
Strongly disagree	1.79%	1.88%	1.84%	2.10%	0.74%
	(1.28–2.45)	(0.89–3.47)	(1.16–2.75)	(0.80–4.43)	(0.09–2.62)
Disagree	1.16%	0.44%	1.18%	3.20%	1.23%
	(0.77–1.68)	(0.05–1.61)	(0.68–1.89)	(1.33–6.39)	(0.25–3.56)
Neither agree nor disagree	7.45%	8.66%	7.32%	6.38%	4.85%
	(6.28–8.75)	(6.28–11.57)	(5.78–9.11)	(3.59–10.35)	(2.49–8.41)
Agree	43.63%	41.13%	44.99%	46.09%	40.92%
	(41.41–45.88)	(36.79–45.58)	(41.94–48.06)	(39.76–52.53)	(33.81–48.32)
Strongly agree	45.97%	47.89%	44.68%	42.22%	52.27%
	(43.72–48.23)	(43.38–52.43)	(41.66–47.74)	(35.87–48.78)	(44.75–59.71)
Teaches how to intervene when you witness a situation involving a fellow Service member (bystander intervention).					
Strongly disagree	1.99%	2.08%	2.06%	2.59%	0.29%
	(1.44–2.67)	(1.02–3.74)	(1.34–3.03)	(1.10–5.10)	(0.00–1.87)
Disagree	1.44%	1.06%	1.24%	2.20%	3.31%
	(0.97–2.05)	(0.31–2.58)	(0.73–1.97)	(0.80–4.74)	(0.96–8.03)
Neither agree nor disagree	8.61%	10.39%	8.17%	8.25%	4.86%
	(7.39–9.95)	(7.81–13.48)	(6.62–9.95)	(5.24–12.23)	(2.40–8.65)
Agree	44.17%	41.78%	44.79%	49.92%	41.72%
	(41.93–46.42)	(37.43–46.24)	(41.73–47.88)	(43.47–56.36)	(34.24–49.50)

Table C.15.c—Continued

	Total	E1–E4	E5–E9	O1–O3	O4–O6
Strongly agree	43.81% (41.57–46.06)	44.69% (40.20–49.24)	43.73% (40.72–46.77)	37.04% (30.80–43.63)	49.82% (42.31–57.33)

Teaches how to obtain medical care following a sexual assault.

	Total	E1–E4	E5–E9	O1–O3	O4–O6
Strongly disagree	1.73% (1.21–2.40)	2.02% (0.93–3.78)	1.57% (0.96–2.42)	2.79% (1.16–5.59)	0.29% (0.00–1.87)
Disagree	2.36% (1.79–3.06)	1.20% (0.55–2.26)	2.50% (1.73–3.50)	4.62% (2.34–8.10)	3.11% (0.81–7.99)
Neither agree nor disagree	9.34% (8.07–10.73)	10.96% (8.33–14.08)	8.77% (7.12–10.65)	9.32% (6.21–13.29)	6.74% (3.92–10.69)
Agree	43.76% (41.53–46.01)	40.12% (35.80–44.55)	45.38% (42.33–48.46)	46.76% (40.39–53.21)	43.10% (35.59–50.85)
Strongly agree	42.81% (40.58–45.06)	45.71% (41.20–50.26)	41.77% (38.78–44.81)	36.51% (30.27–43.10)	46.76% (39.37–54.26)

Explains the role of the chain of command in handling sexual assaults.

	Total	E1–E4	E5–E9	O1–O3	O4–O6
Strongly disagree	1.75% (1.22–2.43)	2.36% (1.19–4.18)	1.56% (0.95–2.41)	2.11% (0.80–4.45)	0.17% (0.00–1.66)
Disagree	1.46% (1.03–2.00)	0.74% (0.28–1.59)	1.70% (1.08–2.54)	2.14% (0.56–5.49)	1.78% (0.53–4.31)
Neither agree nor disagree	7.81% (6.66–9.08)	10.55% (7.99–13.60)	6.80% (5.41–8.42)	8.21% (5.25–12.12)	3.19% (1.31–6.38)
Agree	44.83% (42.59–47.09)	41.38% (37.02–45.85)	46.81% (43.74–49.89)	49.71% (43.26–56.17)	38.49% (31.58–45.76)
Strongly agree	44.15% (41.90–46.41)	44.96% (40.46–49.52)	43.14% (40.12–46.19)	37.83% (31.54–44.42)	56.37% (48.96–63.57)

Explains the reporting options available if a sexual assault occurs.

	Total	E1–E4	E5–E9	O1–O3	O4–O6
Strongly disagree	1.57% (1.07–2.22)	2.05% (0.96–3.81)	1.41% (0.83–2.23)	2.01% (0.73–4.36)	0.17% (0.00–1.67)
Disagree	0.77% (0.47–1.19)	0.28% (0.03–1.04)	1.01% (0.54–1.69)	0.26% (0.01–1.41)	1.78% (0.42–4.82)
Neither agree nor disagree	6.32% (5.27–7.52)	9.02% (6.57–11.99)	5.56% (4.30–7.06)	4.60% (2.52–7.63)	3.02% (1.20–6.19)
Agree	43.73% (41.50–45.98)	41.87% (37.51–46.34)	45.77% (42.72–48.85)	47.52% (41.12–53.99)	31.71% (25.38–38.57)
Strongly agree	47.60% (45.35–49.87)	46.78% (42.27–51.32)	46.25% (43.21–49.31)	45.60% (39.18–52.14)	63.32% (56.19–70.05)

Identifies the points of contact for reporting sexual assault (e.g., SARC, Victim Advocate).

	Total	E1–E4	E5–E9	O1–O3	O4–O6
Strongly disagree	1.65% (1.14–2.32)	2.05% (0.96–3.80)	1.49% (0.89–2.35)	2.41% (0.96–4.95)	0.17% (0.00–1.66)
Disagree	0.93% (0.61–1.37)	0.60% (0.15–1.59)	1.05% (0.59–1.74)	0.53% (0.07–1.84)	1.95% (0.53–4.93)
Neither agree nor disagree	7.35% (6.21–8.61)	9.44% (6.99–12.38)	6.64% (5.24–8.28)	6.00% (3.47–9.57)	5.57% (2.51–10.47)
Agree	43.81% (41.58–46.06)	41.68% (37.32–46.14)	45.60% (42.54–48.68)	47.14% (40.75–53.59)	35.40% (28.76–42.49)
Strongly agree	46.26% (44.01–48.52)	46.23% (41.74–50.78)	45.21% (42.18–48.27)	43.91% (37.53–50.45)	56.91% (49.49–64.11)

Table C.15.c—Continued

	Total	E1–E4	E5–E9	O1–O3	O4–O6
Explains how sexual assault is a mission readiness problem.					
Strongly disagree	1.74%	2.20%	1.59%	2.19%	0.28%
	(1.22–2.40)	(1.09–3.94)	(0.97–2.46)	(0.87–4.50)	(0.00–1.87)
Disagree	1.06%	0.91%	1.01%	2.06%	0.64%
	(0.71–1.51)	(0.35–1.91)	(0.55–1.69)	(0.85–4.17)	(0.06–2.46)
Neither agree nor disagree	8.10%	10.66%	6.82%	8.05%	6.78%
	(6.88–9.46)	(8.04–13.78)	(5.32–8.58)	(4.96–12.21)	(3.44–11.79)
Agree	42.58%	40.94%	43.85%	47.13%	34.17%
	(40.36–44.82)	(36.59–45.39)	(40.82–46.92)	(40.75–53.57)	(27.61–41.22)
Strongly agree	46.53%	45.29%	46.73%	40.57%	58.12%
	(44.28–48.79)	(40.79–49.85)	(43.69–49.79)	(34.24–47.15)	(50.73–65.25)
Explains the resources available to victims (e.g., Safe Helpline).					
Strongly disagree	1.73%	2.10%	1.51%	2.51%	0.68%
	(1.21–2.40)	(1.00–3.86)	(0.91–2.37)	(1.04–5.04)	(0.06–2.74)
Disagree	1.16%	0.99%	1.25%	1.28%	1.13%
	(0.79–1.65)	(0.38–2.08)	(0.73–1.98)	(0.42–2.93)	(0.23–3.31)
Neither agree nor disagree	7.18%	9.09%	6.78%	6.08%	3.78%
	(5.97–8.55)	(6.69–11.99)	(5.10–8.79)	(3.60–9.53)	(1.71–7.13)
Agree	44.06%	40.90%	45.53%	50.34%	38.09%
	(41.84–46.30)	(36.55–45.35)	(42.49–48.60)	(43.88–56.79)	(31.18–45.38)
Strongly agree	45.87%	46.93%	44.93%	39.79%	56.32%
	(43.61–48.13)	(42.41–51.48)	(41.89–48.00)	(33.47–46.37)	(48.89–63.54)

NOTE: Includes estimates for active-component Coast Guard members. 95-percent confidence intervals for each estimate are included in parentheses.

Too few warrant officers were included in the sample to break them out as a separate pay grade. For the purposes of this table, warrant officers have been included with the E5–E9 category.

C.16. Exposure to sexual harassment training

Table C.16.a
Exposure to sexual harassment training, by gender

Longform24: Have you had any military training during the past 12 months on topics related to sexual harassment?

Total	Men	Women
97.90%	98.02%	97.23%
(97.24–98.44)	(97.24–98.63)	(96.15–98.07)

NOTE: Includes estimates for active-component Coast Guard members. 95-percent confidence intervals for each estimate are included in parentheses.

Table C.16.b
Exposure to sexual harassment training, by service

Longform24: Have you had any military training during the past 12 months on topics related to sexual harassment?

Total DoD	Army	Navy	Air Force	Marine Corps	Coast Guard
97.44%	96.52%	98.16%	97.63%	98.35%	97.90%
(96.78–97.99)	(94.79–97.79)	(97.44–98.71)	(97.08–98.11)	(97.30–99.06)	(97.24–98.44)

NOTE: Includes estimates for active-component DoD and Coast Guard members. 95-percent confidence intervals for each estimate are included in parentheses.

Table C.16.c
Exposure to sexual harassment training, by pay grade

Longform24: Have you had any military training during the past 12 months on topics related to sexual harassment?

Total	E1–E4	E5–E9	O1–O3	O4–O6
97.90%	98.85%	98.03%	95.74%	96.05%
(97.24–98.44)	(97.68–99.52)	(97.11–98.72)	(92.77–97.74)	(91.46–98.58)

NOTE: Includes estimates for active-component Coast Guard members. 95-percent confidence intervals for each estimate are included in parentheses.

Too few warrant officers were included in the sample to break them out as a separate pay grade. For the purposes of this table, warrant officers have been included with the E5–E9 category.

C.17. Perception of bystander responsibilities and trust in the military system

Table C.17.a
Perception of bystander responsibilities and trust in the military system, by gender

Longform25a–d: How much do you agree with the following....

	Total	Men	Women
When you are in a social setting, it is your duty to stop a fellow Service member from doing something potentially harmful to themselves or others.			
Strongly disagree	1.15% (0.75–1.69)	1.05% (0.61–1.67)	1.69% (0.88–2.94)
Disagree	0.19% (0.07–0.43)	0.10% (0.01–0.40)	0.71% (0.30–1.40)
Neither agree nor disagree	6.10% (5.01–7.35)	5.98% (4.72–7.44)	6.79% (5.43–8.38)
Agree	30.61% (28.57–32.71)	29.46% (27.11–31.90)	37.09% (34.32–39.92)
Strongly agree	61.95% (59.74–64.13)	63.41% (60.85–65.92)	53.72% (50.81–56.60)
If you are sexually assaulted, you can trust the military system to protect your privacy.			
Strongly disagree	3.49% (2.79–4.32)	3.07% (2.29–4.03)	5.89% (4.62–7.38)
Disagree	8.84% (7.65–10.16)	7.52% (6.19–9.05)	16.32% (14.24–18.58)
Neither agree nor disagree	22.14% (20.32–24.04)	21.02% (18.94–23.22)	28.44% (25.87–31.12)
Agree	38.25% (36.08–40.46)	39.17% (36.66–41.73)	33.03% (30.37–35.77)
Strongly agree	27.28% (25.27–29.36)	29.21% (26.88–31.63)	16.32% (14.21–18.61)
If you are sexually assaulted, you can trust the military system to ensure your safety following the incident.			
Strongly disagree	2.11% (1.56–2.79)	1.86% (1.25–2.66)	3.51% (2.54–4.70)
Disagree	3.28% (2.61–4.07)	2.58% (1.85–3.50)	7.25% (5.86–8.85)
Neither agree nor disagree	17.96% (16.28–19.73)	16.63% (14.72–18.69)	25.43% (22.94–28.05)
Agree	42.71% (40.50–44.94)	42.95% (40.40–45.54)	41.32% (38.50–44.18)
Strongly agree	33.94% (31.80–36.13)	35.97% (33.49–38.50)	22.50% (20.11–25.02)

Table C.17.a—Continued

	Total	Men	Women
If you are sexually assaulted, you can trust the military system to treat you with dignity and respect.			
Strongly disagree	2.63% (2.01–3.36)	2.26% (1.59–3.12)	4.68% (3.53–6.07)
Disagree	4.46% (3.67–5.37)	3.70% (2.82–4.75)	8.81% (7.28–10.55)
Neither agree nor disagree	19.91% (18.17–21.74)	17.76% (15.78–19.87)	32.07% (29.39–34.84)
Agree	41.14% (38.92–43.39)	42.52% (39.95–45.12)	33.38% (30.71–36.13)
Strongly agree	31.86% (29.77–34.00)	33.77% (31.35–36.25)	21.06% (18.72–23.54)

NOTE: Includes estimates for active-component Coast Guard members. 95-percent confidence intervals for each estimate are included in parentheses.

Table C.17.b

Perception of bystander responsibilities and trust in the military system, by service

Longform25a–d: How much do you agree with the following....

	Total DoD	Army	Navy	Air Force	Marine Corps	Coast Guard
When you are in a social setting, it is your duty to stop a fellow Service member from doing something potentially harmful to themselves or others.						
Strongly disagree	1.49% (1.14–1.90)	1.41% (0.94–2.04)	1.74% (0.85–3.14)	0.93% (0.65–1.29)	2.22% (1.19–3.76)	1.15% (0.75–1.69)
Disagree	0.45% (0.30–0.64)	0.30% (0.17–0.48)	0.69% (0.35–1.21)	0.27% (0.13–0.48)	0.76% (0.19–2.04)	0.19% (0.07–0.43)
Neither agree nor disagree	6.92% (6.17–7.73)	9.26% (7.80–10.88)	5.84% (4.66–7.21)	3.94% (3.33–4.61)	7.64% (5.13–10.87)	6.10% (5.01–7.35)
Agree	28.71% (27.41–30.04)	28.00% (25.87–30.21)	31.55% (28.44–34.78)	27.95% (26.50–29.43)	27.05% (22.71–31.74)	30.61% (28.57–32.71)
Strongly agree	62.43% (60.98–63.86)	61.03% (58.62–63.39)	60.19% (56.63–63.67)	66.91% (65.36–68.44)	62.32% (57.39–67.07)	61.95% (59.74–64.13)
If you are sexually assaulted, you can trust the military system to protect your privacy.						
Strongly disagree	4.37% (3.75–5.06)	4.71% (3.51–6.18)	4.29% (3.35–5.41)	3.45% (2.93–4.04)	5.18% (3.16–7.97)	3.49% (2.79–4.32)
Disagree	7.48% (6.71–8.31)	7.17% (6.09–8.37)	8.48% (6.60–10.68)	6.76% (5.89–7.72)	7.85% (5.13–11.40)	8.84% (7.65–10.16)
Neither agree nor disagree	19.84% (18.70–21.03)	22.69% (20.76–24.72)	20.03% (17.19–23.12)	17.34% (16.15–18.58)	16.15% (12.98–19.75)	22.14% (20.32–24.04)
Agree	32.30% (30.86–33.77)	32.47% (30.23–34.77)	35.22% (31.32–39.26)	30.90% (29.43–32.39)	29.25% (25.19–33.58)	38.25% (36.08–40.46)
Strongly agree	36.00% (34.55–37.47)	32.95% (30.64–35.33)	31.98% (28.55–35.56)	41.55% (39.93–43.19)	41.56% (36.70–46.54)	27.28% (25.27–29.36)
If you are sexually assaulted, you can trust the military system to ensure your safety following the incident.						
Strongly disagree	3.00% (2.55–3.50)	3.53% (2.72–4.50)	3.01% (2.17–4.07)	2.07% (1.66–2.54)	3.15% (1.75–5.20)	2.11% (1.56–2.79)
Disagree	4.02% (3.44–4.67)	3.97% (3.27–4.78)	4.26% (2.69–6.39)	3.57% (2.84–4.41)	4.53% (2.78–6.91)	3.28% (2.61–4.07)
Neither agree nor disagree	16.44% (15.45–17.48)	20.35% (18.40–22.40)	15.13% (13.22–17.19)	13.83% (12.75–14.96)	12.68% (10.03–15.72)	17.96% (16.28–19.73)
Agree	35.58% (34.09–37.09)	34.90% (32.61–37.25)	39.09% (35.18–43.11)	33.50% (32.01–35.02)	34.93% (30.25–39.83)	42.71% (40.50–44.94)
Strongly agree	40.96% (39.46–42.47)	37.24% (34.86–39.67)	38.51% (34.77–42.34)	47.04% (45.40–48.68)	44.72% (39.86–49.65)	33.94% (31.80–36.13)

Table C.17.b—Continued

	Total DoD	Army	Navy	Air Force	Marine Corps	Coast Guard
If you are sexually assaulted, you can trust the military system to treat you with dignity and respect.						
Strongly disagree	4.23% (3.51–5.05)	4.50% (3.28–6.02)	4.30% (2.83–6.24)	2.80% (2.30–3.38)	5.86% (3.42–9.27)	2.63% (2.01–3.36)
Disagree	4.03% (3.53–4.58)	3.92% (3.09–4.90)	4.82% (3.61–6.29)	3.75% (3.02–4.59)	3.47% (2.41–4.84)	4.46% (3.67–5.37)
Neither agree nor disagree	18.18% (17.15–19.24)	20.88% (19.04–22.82)	17.41% (15.28–19.71)	15.91% (14.80–17.07)	16.10% (12.88–19.75)	19.91% (18.17–21.74)
Agree	33.33% (31.87–34.83)	33.65% (31.38–35.98)	35.87% (31.97–39.92)	31.57% (30.09–33.07)	31.15% (26.74–35.83)	41.14% (38.92–43.39)
Strongly agree	40.23% (38.74–41.73)	37.04% (34.68–39.46)	37.60% (33.87–41.44)	45.98% (44.34–47.62)	43.43% (38.59–48.35)	31.86% (29.77–34.00)

NOTE: Includes estimates for active-component DoD and Coast Guard members. 95-percent confidence intervals for each estimate are included in parentheses.

Table C.17.c
Perception of bystander responsibilities and trust in the military system, by pay grade

Longform25a–d: How much do you agree with the following....

	Total	E1–E4	E5–E9	O1–O3	O4–O6
When you are in a social setting, it is your duty to stop a fellow Service member from doing something potentially harmful to themselves or others.					
Strongly disagree	1.15% (0.75–1.69)	0.99% (0.36–2.16)	1.23% (0.67–2.06)	1.60% (0.51–3.73)	0.62% (0.04–2.68)
Disagree	0.19% (0.07–0.43)	0.12% (0.00–0.67)	0.23% (0.05–0.64)	0.38% (0.01–2.08)	0.00% (0.00–1.67)
Neither agree nor disagree	6.10% (5.01–7.35)	10.49% (7.69–13.88)	4.71% (3.58–6.08)	2.51% (1.22–4.54)	2.88% (0.99–6.42)
Agree	30.61% (28.57–32.71)	33.66% (29.54–37.98)	30.41% (27.66–33.26)	27.43% (22.33–33.01)	24.02% (17.08–32.15)
Strongly agree	61.95% (59.74–64.13)	54.73% (50.21–59.20)	63.42% (60.46–66.31)	68.09% (62.28–73.50)	72.47% (64.37–79.61)
If you are sexually assaulted, you can trust the military system to protect your privacy.					
Strongly disagree	3.49% (2.79–4.32)	3.27% (1.97–5.08)	3.54% (2.59–4.72)	4.49% (2.50–7.37)	2.73% (1.15–5.40)
Disagree	8.84% (7.65–10.16)	7.82% (5.76–10.32)	8.57% (6.92–10.47)	11.05% (7.68–15.25)	11.88% (7.66–17.32)
Neither agree nor disagree	22.14% (20.32–24.04)	22.74% (19.08–26.74)	22.17% (19.73–24.77)	23.08% (18.08–28.70)	18.19% (13.35–23.90)
Agree	38.25% (36.08–40.46)	38.39% (34.08–42.84)	37.34% (34.45–40.31)	39.70% (33.45–46.21)	42.02% (34.51–49.81)
Strongly agree	27.28% (25.27–29.36)	27.78% (23.79–32.05)	28.37% (25.60–31.26)	21.68% (16.79–27.25)	25.18% (19.31–31.80)
If you are sexually assaulted, you can trust the military system to ensure your safety following the incident.					
Strongly disagree	2.11% (1.56–2.79)	2.49% (1.39–4.08)	2.09% (1.36–3.06)	2.13% (0.84–4.40)	0.67% (0.07–2.50)
Disagree	3.28% (2.61–4.07)	3.60% (2.36–5.24)	3.18% (2.25–4.35)	3.69% (2.04–6.11)	2.20% (0.73–5.02)
Neither agree nor disagree	17.96% (16.28–19.73)	20.10% (16.57–24.02)	18.03% (15.82–20.41)	16.03% (11.99–20.78)	11.41% (7.40–16.59)
Agree	42.71% (40.50–44.94)	42.77% (38.36–47.27)	42.65% (39.65–45.70)	43.69% (37.48–50.06)	41.54% (34.43–48.92)
Strongly agree	33.94% (31.80–36.13)	31.04% (26.93–35.38)	34.04% (31.15–37.02)	34.46% (28.30–41.03)	44.18% (36.67–51.88)

Table C.17.c—Continued

	Total	E1–E4	E5–E9	O1–O3	O4–O6
If you are sexually assaulted, you can trust the military system to treat you with dignity and respect.					
Strongly disagree	2.63% (2.01–3.36)	2.94% (1.77–4.58)	2.54% (1.73–3.61)	3.10% (1.51–5.59)	1.30% (0.32–3.46)
Disagree	4.46% (3.67–5.37)	4.27% (2.85–6.13)	4.56% (3.46–5.88)	5.15% (3.00–8.17)	3.68% (1.74–6.76)
Neither agree nor disagree	19.91% (18.17–21.74)	23.02% (19.39–26.98)	19.12% (16.83–21.58)	19.07% (14.86–23.88)	13.99% (9.32–19.87)
Agree	41.14% (38.92–43.39)	39.12% (34.76–43.61)	40.84% (37.84–43.89)	44.08% (37.69–50.62)	47.38% (39.87–54.98)
Strongly agree	31.86% (29.77–34.00)	30.64% (26.55–34.97)	32.94% (30.07–35.90)	28.60% (23.16–34.55)	33.65% (27.07–40.73)

NOTE: Includes estimates for active-component Coast Guard members. 95-percent confidence intervals for each estimate are included in parentheses.

Too few warrant officers were included in the sample to break them out as a separate pay grade. For the purposes of this table, warrant officers have been included with the E5–E9 category.

C.18. Perceived trends in sexual harassment and sexual assault

Table C.18.a
Perceived trends in sexual harassment and sexual assault, by gender

Longform 26–29

	Total	Men	Women
In your opinion, has sexual harassment in our nation become more or less of a problem over the last 2 years?			
Less of a problem today	21.87% (20.01–23.82)	23.22% (21.06–25.49)	14.30% (12.30–16.50)
About the same as 2 years ago	35.36% (33.27–37.50)	34.56% (32.15–37.03)	39.91% (37.13–42.74)
More of a problem today	22.41% (20.57–24.33)	21.71% (19.60–23.94)	26.35% (23.82–29.01)
Do not know	20.35% (18.55–22.25)	20.52% (18.43–22.72)	19.43% (17.25–21.76)
In your opinion, has sexual harassment in the military become more or less of a problem over the last 2 years?			
Less of a problem today	31.54% (29.43–33.71)	33.46% (31.02–35.98)	20.71% (18.39–23.18)
About the same as 2 years ago	34.52% (32.44–36.65)	33.63% (31.23–36.09)	39.53% (36.73–42.37)
More of a problem today	16.41% (14.74–18.19)	16.01% (14.09–18.07)	18.70% (16.46–21.11)
Do not know	17.53% (15.83–19.33)	16.90% (14.95–18.99)	21.06% (18.77–23.50)
In your opinion, has sexual assault in our nation become more or less of a problem over the last 2 years?			
Less of a problem today	18.20% (16.43–20.08)	19.71% (17.65–21.91)	9.71% (7.98–11.67)
About the same as 2 years ago	38.05% (35.91–40.23)	37.54% (35.07–40.06)	40.95% (38.14–43.80)
More of a problem today	23.03% (21.13–25.00)	22.25% (20.07–24.55)	27.39% (24.85–30.04)
Do not know	20.72% (18.97–22.55)	20.50% (18.49–22.62)	21.95% (19.65–24.39)

Table C.18.a—Continued

	Total	Men	Women
In your opinion, has sexual assault in the military become more or less of a problem over the last 2 years?			
Less of a problem today	30.36% (28.28–32.49)	32.19% (29.78–34.66)	19.98% (17.69–22.44)
About the same as 2 years ago	34.35% (32.27–36.47)	33.32% (30.93–35.77)	40.18% (37.37–43.05)
More of a problem today	16.14% (14.52–17.88)	15.75% (13.88–17.76)	18.40% (16.21–20.77)
Do not know	19.15% (17.37–21.03)	18.75% (16.70–20.94)	21.43% (19.14–23.87)

NOTE: Includes estimates for active-component Coast Guard members.
95-percent confidence intervals for each estimate are included in parentheses.

Table C.18.b

Perceived trends in sexual harassment and sexual assault, by service

Longform26–29

	Total DoD	Army	Navy	Air Force	Marine Corps	Coast Guard
In your opinion, has sexual harassment in our nation become more or less of a problem over the last 2 years?						
Less of a problem today	19.38% (18.10–20.71)	18.81% (16.76–21.00)	21.93% (18.71–25.42)	16.93% (15.69–18.22)	20.75% (16.86–25.08)	21.87% (20.01–23.82)
About the same as 2 years ago	31.56% (30.28–32.87)	31.33% (29.18–33.54)	29.52% (26.79–32.36)	34.43% (32.92–35.97)	30.77% (26.33–35.49)	35.36% (33.27–37.50)
More of a problem today	26.09% (24.69–27.53)	29.14% (27.00–31.35)	25.35% (21.46–29.56)	23.72% (22.35–25.13)	23.23% (19.36–27.47)	22.41% (20.57–24.33)
Do not know	22.97% (21.69–24.28)	20.72% (18.74–22.82)	23.20% (20.00–26.65)	24.92% (23.49–26.39)	25.25% (21.22–29.62)	20.35% (18.55–22.25)
In your opinion, has sexual harassment in the military become more or less of a problem over the last 2 years?						
Less of a problem today	28.98% (27.58–30.40)	27.27% (25.03–29.60)	31.68% (28.21–35.30)	28.41% (26.94–29.92)	29.87% (25.36–34.69)	31.54% (29.43–33.71)
About the same as 2 years ago	28.81% (27.58–30.07)	29.93% (27.78–32.14)	26.54% (23.97–29.24)	30.35% (28.89–31.85)	27.08% (23.07–31.39)	34.52% (32.44–36.65)
More of a problem today	21.33% (19.97–22.74)	24.12% (22.08–26.25)	20.18% (16.38–24.42)	18.87% (17.59–20.21)	20.09% (16.20–24.44)	16.41% (14.74–18.19)
Do not know	20.88% (19.62–22.19)	18.68% (16.74–20.75)	21.60% (18.34–25.16)	22.36% (21.00–23.77)	22.96% (19.08–27.21)	17.53% (15.83–19.33)
In your opinion, has sexual assault in our nation become more or less of a problem over the last 2 years?						
Less of a problem today	16.17% (14.92–17.49)	14.92% (13.04–16.96)	18.80% (15.43–22.56)	14.58% (13.42–15.81)	17.77% (14.02–22.05)	18.20% (16.43–20.08)
About the same as 2 years ago	33.83% (32.51–35.17)	35.56% (33.29–37.88)	30.32% (27.55–33.21)	35.39% (33.85–36.95)	32.49% (28.14–37.08)	38.05% (35.91–40.23)
More of a problem today	24.12% (22.82–25.46)	26.28% (24.27–28.37)	23.91% (20.23–27.90)	22.85% (21.50–24.25)	20.85% (17.66–24.33)	23.03% (21.13–25.00)
Do not know	25.88% (24.53–27.26)	23.24% (21.17–25.42)	26.97% (23.65–30.49)	27.17% (25.71–28.68)	28.89% (24.26–33.86)	20.72% (18.97–22.55)
In your opinion, has sexual assault in the military become more or less of a problem over the last 2 years?						
Less of a problem today	28.33% (26.92–29.77)	26.15% (23.93–28.48)	31.34% (27.76–35.10)	27.51% (26.04–29.01)	30.45% (25.97–35.22)	30.36% (28.28–32.49)
About the same as 2 years ago	29.23% (27.99–30.50)	30.77% (28.63–32.97)	26.82% (24.19–29.58)	30.35% (28.89–31.84)	27.31% (23.16–31.78)	34.35% (32.27–36.47)
More of a problem today	20.28% (18.99–21.63)	22.62% (20.69–24.65)	18.35% (14.84–22.28)	19.00% (17.73–20.33)	19.53% (15.54–24.03)	16.14% (14.52–17.88)
Do not know	22.15% (20.87–23.47)	20.45% (18.41–22.62)	23.49% (20.16–27.08)	23.14% (21.76–24.56)	22.71% (19.08–26.67)	19.15% (17.37–21.03)

NOTE: Includes estimates for active-component DoD and Coast Guard members. 95-percent confidence intervals for each estimate are included in parentheses.

Table C.18.c
Perceived trends in sexual harassment and sexual assault, by pay grade

Longform 26–29

	Total	E1–E4	E5–E9	O1–O3	O4–O6
In your opinion, has sexual harassment in our nation become more or less of a problem over the last 2 years?					
Less of a problem today	21.87% (20.01–23.82)	22.83% (19.11–26.89)	19.54% (17.23–22.02)	27.75% (21.82–34.31)	26.34% (19.39–34.28)
About the same as 2 years ago	35.36% (33.27–37.50)	29.79% (25.80–34.03)	35.83% (32.97–38.76)	38.65% (32.81–44.74)	50.04% (42.55–57.53)
More of a problem today	22.41% (20.57–24.33)	23.84% (20.20–27.79)	23.74% (21.20–26.44)	17.45% (12.93–22.77)	14.09% (9.71–19.49)
Do not know	20.35% (18.55–22.25)	23.54% (19.89–27.51)	20.88% (18.40–23.54)	16.16% (11.87–21.25)	9.53% (5.96–14.27)
In your opinion, has sexual harassment in the military become more or less of a problem over the last 2 years?					
Less of a problem today	31.54% (29.43–33.71)	30.13% (26.03–34.49)	29.10% (26.33–31.98)	38.13% (31.79–44.79)	45.32% (37.77–53.04)
About the same as 2 years ago	34.52% (32.44–36.65)	28.44% (24.49–32.66)	36.68% (33.82–39.62)	39.17% (33.22–45.36)	37.80% (30.79–45.20)
More of a problem today	16.41% (14.74–18.19)	19.29% (15.84–23.13)	17.18% (14.92–19.64)	10.25% (6.62–14.97)	7.70% (4.44–12.27)
Do not know	17.53% (15.83–19.33)	22.13% (18.65–25.94)	17.04% (14.71–19.57)	12.46% (8.93–16.75)	9.18% (5.64–13.92)
In your opinion, has sexual assault in our nation become more or less of a problem over the last 2 years?					
Less of a problem today	18.20% (16.43–20.08)	19.86% (16.31–23.80)	17.35% (15.11–19.77)	20.98% (15.40–27.50)	13.73% (7.70–22.00)
About the same as 2 years ago	38.05% (35.91–40.23)	32.62% (28.49–36.95)	37.23% (34.33–40.20)	44.98% (38.76–51.32)	56.31% (48.56–63.84)
More of a problem today	23.03% (21.13–25.00)	23.92% (20.24–27.92)	24.52% (21.86–27.34)	16.64% (12.50–21.49)	17.60% (12.50–23.74)
Do not know	20.72% (18.97–22.55)	23.60% (20.00–27.52)	20.90% (18.56–23.39)	17.40% (13.01–22.53)	12.35% (8.27–17.50)
In your opinion, has sexual assault in the military become more or less of a problem over the last 2 years?					
Less of a problem today	30.36% (28.28–32.49)	29.55% (25.48–33.87)	28.24% (25.53–31.07)	37.07% (30.77–43.71)	39.20% (31.79–46.99)
About the same as 2 years ago	34.35% (32.27–36.47)	27.97% (24.02–32.19)	35.73% (32.90–38.64)	40.25% (34.22–46.52)	42.39% (35.21–49.83)
More of a problem today	16.14% (14.52–17.88)	17.49% (14.23–21.15)	17.41% (15.15–19.86)	10.27% (6.92–14.52)	9.89% (6.17–14.81)
Do not know	19.15% (17.37–21.03)	24.99% (21.29–28.99)	18.62% (16.17–21.28)	12.41% (8.93–16.65)	8.52% (5.19–13.02)

NOTE: Includes estimates for active-component Coast Guard members. 95-percent confidence intervals for each estimate are included in parentheses.